I GROW CHRONIC: The Original Step By Step Beginners Guide to the Construction and Maintenance of an Indoor Hydroponic Growroom that will Produce Over 1/2 Pound of Chronic Cannabis Bud Every Two Months.
By Mr. Just Dank Green

Mr. Green's Chronic Entertainment Company, LLC.
www.igrowchronic.com

ISBN- : 978-0-6926-5728-7
Published by Mr. Green's Chronic Entertainment Company, LLC.
Los Angeles, CA.

Created in the United States of America.

CONTENTS

Before we get started let's get one thing straight...

I LOVE GROWING CHRONIC.

Growing cannabis in the great outdoors can be fun and rewarding BUT I really like the higher quality bud that can be achieved when growing indoors. That's what this book is all about. I am going to construct an indoor growroom that will house both a vegetative room (18 hours of light per day) and a flowering room (12 hours of light per day). I'm building two growrooms in one, in order to be self-sufficient. With my independence, I will not need to go to outside sources for my clones. I will build the rooms without help from anyone and maintain them with ease. In the vegetative area, I'll grow mother plants from seed and raise little baby clones for flower production. The flowering (budding) room will have a powerful light with an orange glow which will help me produce around a 1/2 pound of hydro-chronic, medical grade, sinsemilla,

A+ CANNABIS!

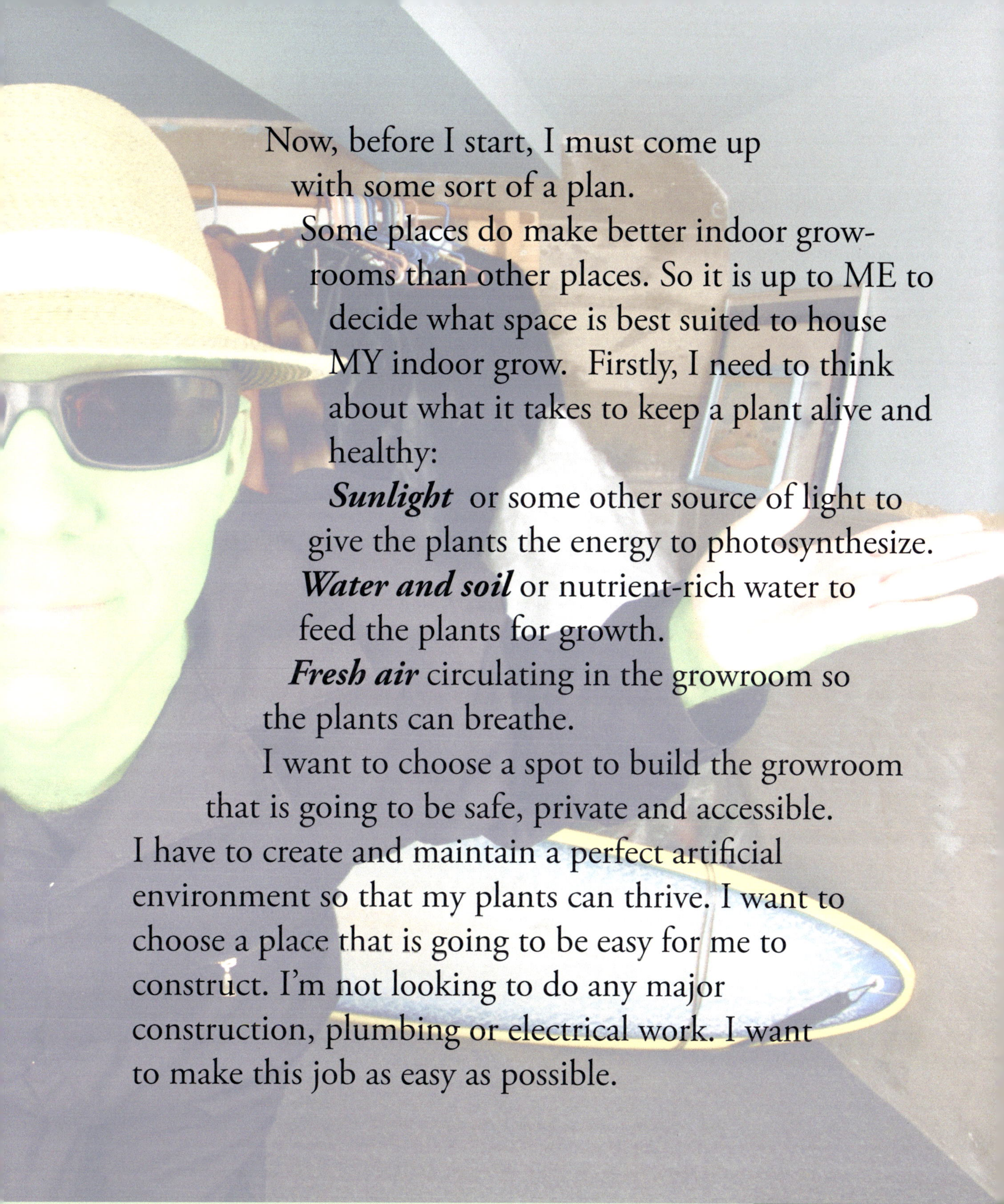

Now, before I start, I must come up with some sort of a plan.

Some places do make better indoor grow-rooms than other places. So it is up to ME to decide what space is best suited to house MY indoor grow. Firstly, I need to think about what it takes to keep a plant alive and healthy:

Sunlight or some other source of light to give the plants the energy to photosynthesize.

Water and soil or nutrient-rich water to feed the plants for growth.

Fresh air circulating in the growroom so the plants can breathe.

I want to choose a spot to build the growroom that is going to be safe, private and accessible. I have to create and maintain a perfect artificial environment so that my plants can thrive. I want to choose a place that is going to be easy for me to construct. I'm not looking to do any major construction, plumbing or electrical work. I want to make this job as easy as possible.

When building this growroom, I'm looking to create an environment that is perfect for healthy, growing plants. I'm actually going to build two small growrooms so that I can have both a "summer" room and a "fall" room. The "summer" side, vegetative room, is used to grow the plants big and strong and the "fall" side, the flowering room, is used to bud the plants out. I do this two room set-up so I can have plants budding in the flowering room while, at the same time, the next generation is growing in the vegetative room. Having two rooms also eliminates my dependence on other people, as I have a place for the mother plant(s) and clones to grow in an endless summer on the vegetative side while other plants are budding on the flowering side. Both the summer side (vegetative side) and the fall side (flowering / budding room) are going to be about the same size, 4 feet by 5 feet with an 8 foot ceiling. The main difference between the rooms will be in the lighting. The photoperiod will be different for each room as the vegetative room will have the lights turned on for 18 hours a day and the flowering room will have its light on for 12 hours per day. It is the number of hours of light during the day, above all other factors, that triggers the plants to start budding. Also, the lights themselves differ from room to room. The vegetative room will have a few fluorescent lights to keep the plants growing, while the flowering room will have a more powerful 600watt High Pressure Sodium light to support big bud development. The color the light bulbs emit will differ as well. The fluorescents will put off a blue light and the HPS bulbs give off a deep orange color. The flowering room uses the more powerful HPS light as it illuminates the plants during their magical transformation from little green weeds into chronic trees covered with juicy, stinky, skunk buds.

When thinking about where to build the indoor growroom, keep these things in mind...

FRESH AIR FLOW

Plants breathe. Just as we take in oxygen, plants breathe carbon dioxide (CO2). Having fresh air flow in a growroom usually means having some sort of air vents or windows to get air in and out of the room. A closed space will have major issues with heat and humidity! It is essential to have good air flow if I want to create a successful growing environment. When the air vent system is all up and running, I will be able to just kick-it in the room and chill with the plants. It's relaxing to exchange air with them... trading your CO2 for a little of their pure oxygen.

LIGHTING

In nature, cannabis plants need light. In the process called photosynthesis, plants convert light into energy that they use to grow and flower. In nature plants get that light from the sun. If I'm to successfully create an environment that will support plant life, then I too, like nature, must bring in light. There are so many different kinds of lighting systems to choose from because there are many different styles and sizes of indoor growrooms. Lighting is the largest factor in determining the overall size of the bud. I can have great air flow and the best water and nutrients but with a weak light I will only get light, airy buds at best.

WATER & NUTRIENTS

In the wild, plants get their water from the rain and their nutrients from the soil. In the growroom plants get their water from a garden hose and their nutrients from a bottle. I'm not just trying to mimic nature in the growroom, I'm attempting to copy the most positive aspects of nature in regards to plant health. Some of the best things for a plants health is a consistent watering schedule and top quality nutrients. The goal is for the farmer to do a great job and to replicate the best nature can offer in search of amazing results.

More details on these main environmental factors on pages 5 and 6.

ELECTRICITY

In order to have lights, vents, pumps and fans in my growroom I must have a good, safe source of electricity. Fact is, I'm no electrician. I'm not really good at electrical things. All I really need to know is, "can I power the grow or not?" Is my home's circuit breaker box in good working order or is it a piece of junk that is already overloaded, falling apart and ready to burn the whole house down? What kind of wiring and outlets does the space have in it? Is it 120 volt or 240 volt? Is it modern, up to code, romex wire or that old exposed knob and tube wire, dating back to the 1930's? I try to determine the working order of a home's electrical system by examining the breaker box. It's important to know the size of breaker for the grow area. Is it 15 amps or 20 amps? By just looking at the label on the breaker, I can see if the room has a 15amp or a 20amp breaker. If the room has a 20amp+ breaker, then I have nothing to worry about because that is plenty of power for this grow. If the room has only a 15 amp breaker, then I need to be aware of the power demands, as not to overload the wires.

Here is an electrical rule of thumb:
15amp=1500 watt max and 1200 watt LCL.
Long Continiuious Load (LCL).
20amp=2000watt max and 1600 watt max LCL.

15amp breaker can take a maximum load of 1500 watts before tripping off. The 15amp should only take a LCL, Long Continuious Load, of 1200 watts. Growers should use LCL because our equiptment is turned ON for many hours at a time. If I add up the watts of the long running grow equipment, it should total less than 1200watts and if it spikes over 1500watts the breaker will trip.
The 20amp breaker can take 2000watts before tripping and a LCL of 1600watts.
The bigger the breaker the better off the growroom.

I add up how much electricity is going to be used: ELECTRICAL GROW EQUIPMENT
-600watt HPS light (Use 400watt for less power)
-2X's T5 fluorescent lights 384watt
(Use 1x's T5 for less power @192watt)
-6" exit vent fan 130watt. Flowering room.
-4" exit vent fan 98watt. Vegetative room.
-2X's room fans 40watt. Both rooms.
That's 1252 watts of power to grow over 1/2 pound of cannabis flowers!
(-392watts = 860watts for the less powerful lighting options).
This room has one 20 amp curcuit and that will be enough to power this growroom 100%.
If I overload the breaker by trying to pull too much energy, the breaker will trip. The good thing about this size of grow is that it doesn't require a very high demand for energy. Most places have a 20 amp curcuit available, which is enough to run a 600 watt HPS for the flowering room, two T5's for the vegetative room and some fans to move the air around.
(If you want to know more about electricity, try a local tech school and sign up. It is an excellent career option.)

WORD.
A project this size will raise the electricity bill approximately $100 per month.

IF ALL ELSE FAILS.
Hire a professional electrician to bring the future project area up to electrical code.

THE THREE MAIN FACTORS USED TO CREATE THE

AIR FLOW

As before mentioned, plants breathe. So I cannot build a growroom in a closed space without first figuring out how I'm going to get some fresh air into that space. Plants will stop growing in stale air. I must get the air moving. As a rule, it is easier to pull air into a space rather than to push air into a space. By exiting the air in a room with a vent fan I can create a vacuum effect in a room. When air is taken out of a room via vent fan, fresh air will automatically be drawn back into the room through door jams, windows or air vents. Air goes where it is needed. So for this project I will use a 4 inch vent fan attached to a small carbon filter for the vegetative room and a 6 inch vent fan attached to a large carbon filter for the flowering room. I'll connect the vent fans to the carbon filters so that stinky air from the growroom will be filtered as it is sent outside. Fresh air will be drawn into the rooms from some air vents and circulated around the rooms with small fans.

It's true that plants can breathe higher levels of CO_2 than what is available in normal outdoor air. These high levels of CO_2, 1300 parts per million, can produce up to 25% more weight on the plant BUT there are many challenges with operating a successful CO_2 enrichment program. This is a more advanced technique. It will cost a lot ($1500+) more money. It is challenging and more costly to maintain. It will fail and ruin the whole harvest at least once. A CO_2 enrichment program is a better choice for advanced growers.

WATER & NUTRIENTS

The purity of the water and the quality of the nutrients can make-it or break-it on the judgement scale. Attention to detail is what sets some cannabis apart from the rest. As far as the growroom is concerned, it's nice to have easy access to clean water. The more clean and filtered the water is, the better. A project of this size is expected to use maybe 20 gallons of water per week and a sink with a drain would come in useful. As far as nutrients are concerned, I use top-of-the line, liquid, natural (from natual sources) nutrients to ensure maximum growth and flowering potential. What that means is... I mix powerful nutrients into water and pour this mix over the base of the plants to feed them.

The roots at the base of the plant grow in a soil-less block cube. Hydroponic means grown without soil and by definition cannot be organic. Nevertheless, I try to grow as natural as possible and avoid harsh chemicals. I use excellent nutrients to achieve healthy growth rates and intense herbal flavor. When growing in hydroponics, I can get a ton of trichome crystals on my flowers without having to move a ton of soil around the house.

LIGHTING

To ensure I grow at least a 1/2 pound of solid bud, I will power my flower room with a 600watt High Pressure Sodium light with electric ballast. I've had success in the past with a 400watt light but the amount and quality of cannabis I can produce with a 600watt is more solid. A 1000watt HPS is a great choice if I was looking to grow up to and over one pound! But my current cannabis needs are less, so 8 ounces will be fine for my needs. Plus, the overall cost will be lower and the grow will be less stressful to run and maintain. During the 12 hours of daily light needed to bud the plants, the light fixtures can get very hot and thus cause an entire room to overheat. For this reason I'll use an enclosed reflective hood so that the light can be air cooled by connecting it to an air-vent system. This is how the pros do it, because it's easy to set up and it works well. I could run a cheaper lighting system but at what expense? If I did run the flower room without the air cooled hood and the room started to overheat, above 90° F., then what? No herb for Mr. Green! That's what! The more ways I can control the growroom's environment, the better. The HPS light is the greatest source of heat for the growroom, and venting that hot air out of the room is a great way to limit and control that heat. This is also the reason I use an electric ballast. Not only does it have a longer life, it uses up to 40% less energy than the conventional ballast and it puts off 50% less heat!

On the other side is the vegetative room. It has a more simple lighting requirement. Here is where I'll have fluorescent lights turned on for 18 hours per day. I'm going to make sure the bulbs I use emit a light from the blue spectrum.

Grow bulbs = blue light.

Flowering bulbs = orange light.

Fluorescent lights emit less heat, therefore they do not need to be air cooled. Due to the low operating temperature of the bulbs, I can move the lights to within 4 inches of the tops of the plants. Now, I could use any old four foot shop light with some cool-white & warm-white fluorescent bulbs BUT I'm looking to find a more balanced light spectrum specifically designed for growing plants. The compact 2'x2' T5 fluorescent grow lights are the best choice for a small vegetative growroom. With the T5's and the 600watt cranking out the lumens this garden will be ready for action.

LOCATION, LOCATION, LOCATION.

I can build a growroom almost anywhere...

BIG CLOSET

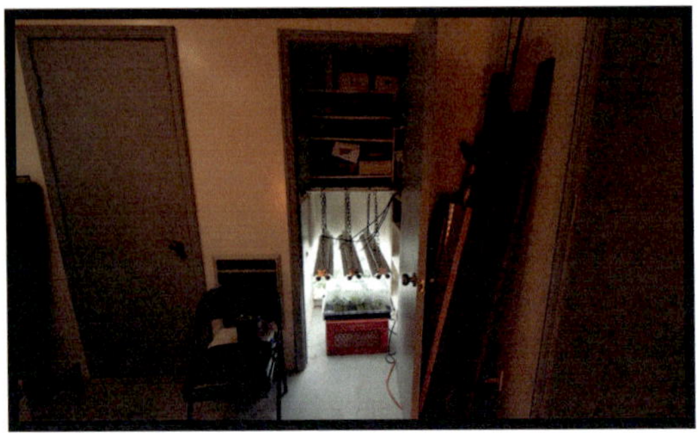

PROS

-Private, secure area inside the house.

-Hidden and out of sight.

CONS

-Vent and air-flow issues are the big challenge in here.

-Tight space. Not much room to work.

OUTDOOR SHED

PROS

-Plenty of space to work.

-Perfect self-enclosed grow zone.

CONS

-Getting electricty and running water.

-More exposed to the outdoor elements and temperatures.

-Security.

RV

PROS

-This ToyHauler has water and electricty.

-It can be a good spot if you have secure parking.

CONS

-May get HOT in the summer sun.

-Lose the use of the RV for camping trips!

GARAGE

PROS

-The garage can be a great location depending on its access to water and power.

CONS

-May be a little "in your face" for the neighborhood.
-Lose the indoor parking spots.

GUEST BEDROOM

PROS

-Most bedrooms have electricty.
-It is nice and secure to have the growroom in the house.

CONS

-Giving up a bedroom.
-Wall to wall carpet, holes in nice walls.

GROW TENT

PROS

-A fabric tent growroom inside an empty spare room is quick and easy to set-up.

CONS

-They can be expensive and have been known to rip apart.

but...

my favorite place to grow is right here...

in the basement.

First things first. I must move this stuff out of here!

CLEAR IT OUT!

CLEAN IT UP!

This carpet is not installed, ie. stapled into the floor. Therefore, it is easy for me to roll it up for storage. When I'm working on installed carpet that I don't want to damage, I'll first put down plastic, then some 1 inch plywood as to protect the carpet and provide me with a stable floor to work on.

I take measurements of the space and see what I have to work with...

The basement has a lot of workable space with 10 feet of width x 8.5 feet height x 4 feet depth. That is plenty of room to work with for both the vegetative and flowering rooms. Now all I've got to do is build it and this room will be ready for action.

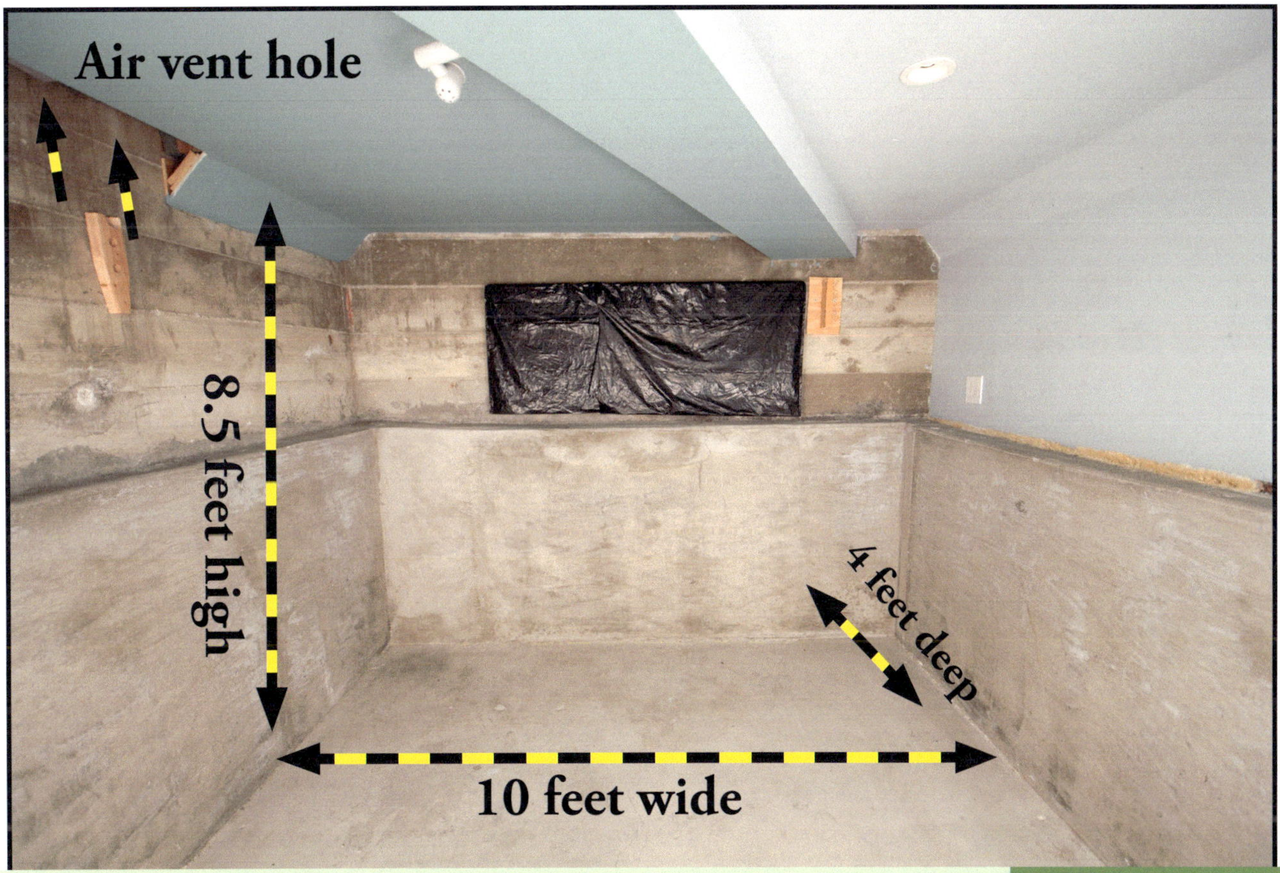

This side will be my flowering room.

This is where I will bud my plants by creating an environment that mimics the short 12 hour days of the late fall season. 12 hours of light and 12 hours of complete darkness is the photoperiod that triggers the plants to go into flowering mode. I must separate this side of the room from the other side so that NO LIGHT from the other, 'vegetative', side will interrupt the sleep of the flowering plants on this side, thus preventing them from budding. I'll do that by building a wood frame and then wrapping it in thick 6mm black plastic. To light this room and bud my plants I'll be using a 600 watt, high pressure sodium, with an electric ballast. I think the powerful 600 watt HPS is a great choice for this spot.

A 1000 watt may be tempting but it increases the costs of everything involved. Remember, all I'm looking to grow is a 1/2 pound every two months and the 1000 watt will give me a whole pound. I've used a 400 watt HPS light in the past, with success, BUT for this project I was looking to grow a more solid, phat bud that only a more powerful light can provide. All that chronic bud growing in the basement is going to stink, big time. I will use a heavy duty carbon filter so that my neighbors cannot smell the skunk flowers and know what's up. The simple ventilation system I'm going to set up will filter the stinky air on the way out of the growroom while bringing fresh air into the room for the plants to breathe. I'm only planning on budding 4 plants for this cycle but the 4' x 5' room will be big enough for 1 or 2 more plants, if I wanted. The girls will be growing inside soilless grow cubes which will be placed inside some hard plastic trays. Perfect.

This side will be my vegetative room.

This is the room that will imitate the 18 hour long days of summer. I plan to have my mother plants and my cloned, preflower production plants growing up healthy over here under some two foot long T5 fluorescent lights.

This side will always have 18 hours of blue spectrum lighting to promote rapid plant growth. Although the smell from the plants in the vegetative room is not nearly as strong as the stink coming from the flowering side, I still use a small carbon filter to keep the smell of the green down. I attach a 4 inch fan to the carbon filter which acts to pull fresh air into the space while pushing cleaned, odorless air to the outside.

I plan on having the mother plants growing up in here, sitting in hydro-air-o-ponic buckets right on the floor. I'll have a 2'x2' plastic tray set-up right next to the mother plants for the clones. Excluding the mother plants who will be living in the bucket systems, I will be watering all the plants by hand so I don't need to worry about extra pumps and timers for feeding. Excellent! This is all I need to be a completely self sufficient grower! Now, let's build it...

Dividing the space into two rooms starts with this center 2"x 4" beam. I first measure and cut the wood beam to fit the room from floor to ceiling. Then I use a L-bracket and screws to secure the beam to the ceiling. From this point I start framing the room out, piece by piece, until I have a complete room, framed up.

What I'm doing here is building the frame for the flowering room. I want to make it strong enough so it will not fall apart and give me problems later.

$ MORE MONEY THAN TIME $

I could go purchase a pre-fabricated grow tent or box for $500+. There are several types on the market. But remember to get one for growing and one for budding.

TOOLS:

ELECTRIC DRILL; DRILL BITS; SCREW DRIVER BIT; ELECTRIC WOOD SAW; TAPE MEASURE.

Okay, now every space is going to have its own unique set of challenges and this room is no different. After framing the flowering room out I've noticed some holes that will need to be sealed up. I'll use some thin drywall pieces to coverup the holes and provide support for the 4 inch and 6 inch vent tubes used to air the room out. I can use the room's pre-existing concrete ledge as a work shelve for the vent tube, electric ballast and carbon filter.

The exit vent hole sits just out of the framed-up flowering room and must be connected to the exit vents. I'll do this by cutting out two holes, a 6 inch and a 4 inch, and attaching vent connectors to them. NOW... I've got this piece of drywall with two perfect vent attachments ready for business. I'll tape the edges with Gorilla Tape ™ (duct tape) to prevent damage and then secure this piece of drywall over the hole. To the next hole...

TOOLS: ELECTRIC DRILL; DRILL BITS; TAPE MEASURE; DUCT TAPE; DRYWALL SAW.

This hole is between the flowering room and the rest of the basement and must be closed. I take some measurements and then cut out a small piece of drywall so that it will fit over and close-up the open space.

Notice I've cut two holes in the piece of drywall so that I can pull air through the exit vent hoses from both rooms to the outside. Once the drywall has been secured to the wood frame, we're ready for the next step.

This last opening sits between the two rooms and must be closed-up in order to prevent light from the vegetative room from entering into and disturbing the sleeping plants in the flowering room. I first measure out the hole's unique dimensions and then cut out a small piece of drywall to cover the hole.
I'll cut out one 4 inch hole to pass the vent hose through. I'll be exiting the air from the vegetative room through this hole. I make sure it fits snug before I drill the drywall into place. The hole is primed and ready for me to pass the four inch vent tube through when it's time to connect the exit / exhaust fan.

MATERALS: 3 small pieces drywall; box of 1" & 3" drywall screws; 4" & 6" vent connectors.

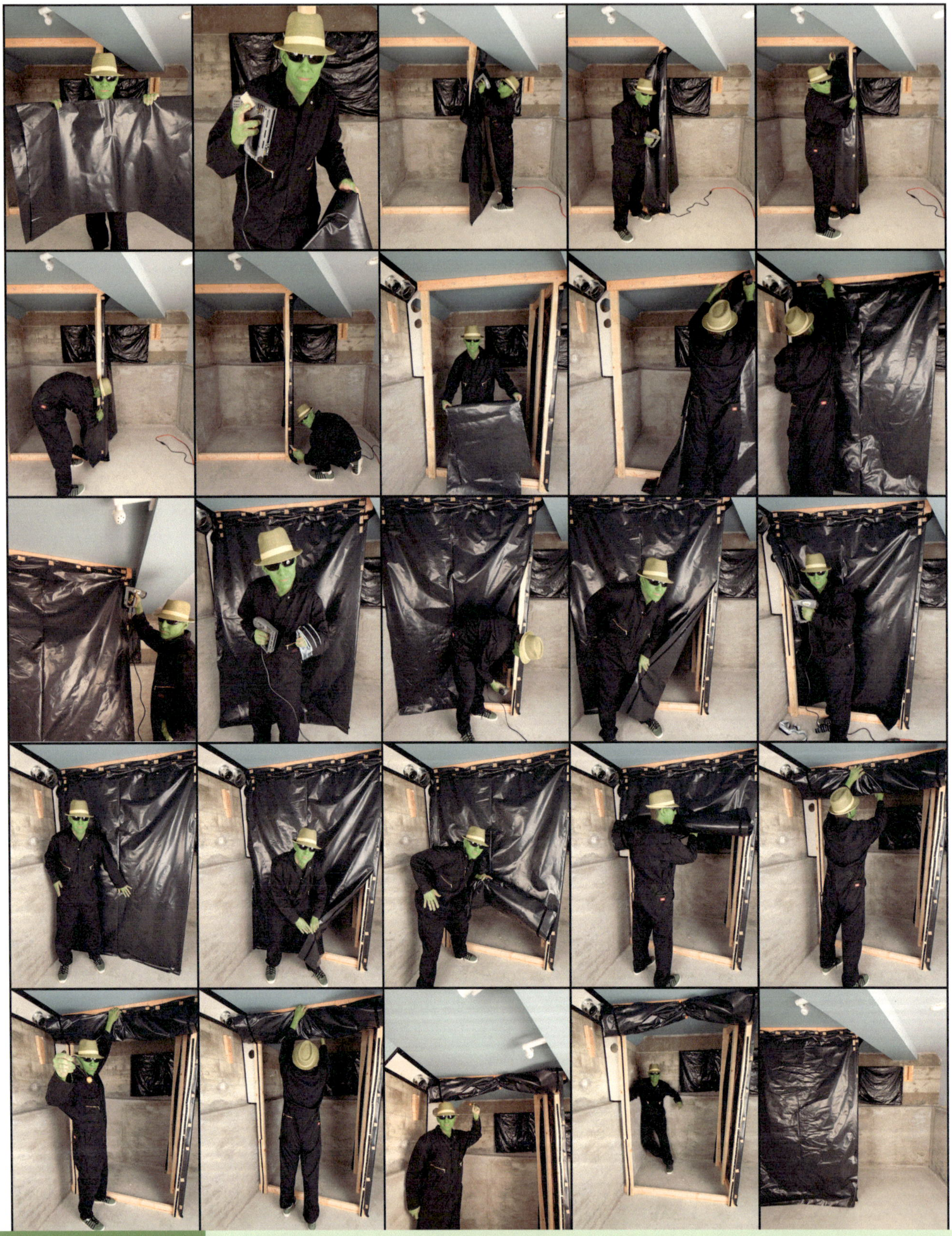

Now that the room is framed-out and the holes are closed up, it's time to cover it all up with black plastic. I use the 6mm thick black plastic which is a little more costly, $$, but a lot more heavy duty and effective in keeping unwanted light out. Now that I've got this new electric stapler, it's much more easy to put up the black plastic. It's important to usc small 1 inch x 1 inch pieces of thin cardboard to staple up, over the plastic. This gives the staple something added to hold onto, thus preventing it from ripping off. Once the plastic is secured up to the frame, I take a razor blade and cut off the undesired excess. Perfect.

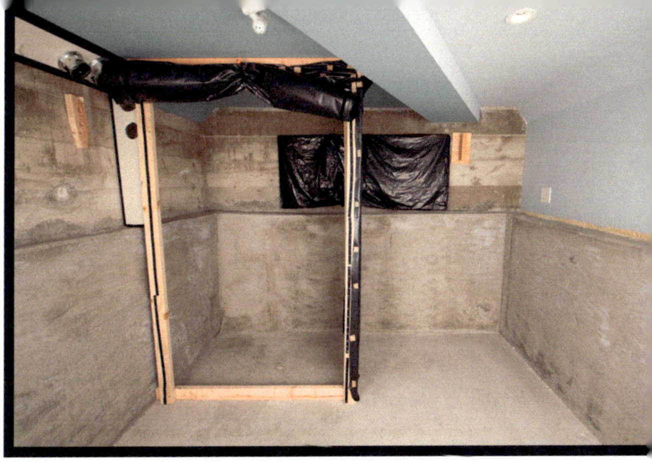

With the wall between the vegetative and flowering rooms sealed off, I focus my attention on the roll-down door. I want this door to be easy to open and close, plus it needs to be light proof. The nice part is that it does not get much easier than setting up a basic roll-down door. First, I cut a big square of plastic that is 100% going to cover the door. Then, I staple it up to the top beam using my staple gun and the tiny cardboard pieces for support. I've put some hooks up on both sides of the top beam so that when I roll the black plastic door up from the bottom, I can use a simple bungee cord to strap the door up in the open position. Totally light proofing the room takes just a little more work. The main tool I use in keeping the plastic door shut is Velcro™. I'll Velcro™ the sides of the plastic to the wood frame so as to prevent the door from opening and letting light inside when rolled down. I check to see if the flowering room is light proof by going inside the space with the door rolled down and examining it for any beams of light that may be shining inward. It needs to be pitch black in there, for 12 continuous hours, in order for the plants to stay in flowering mode.

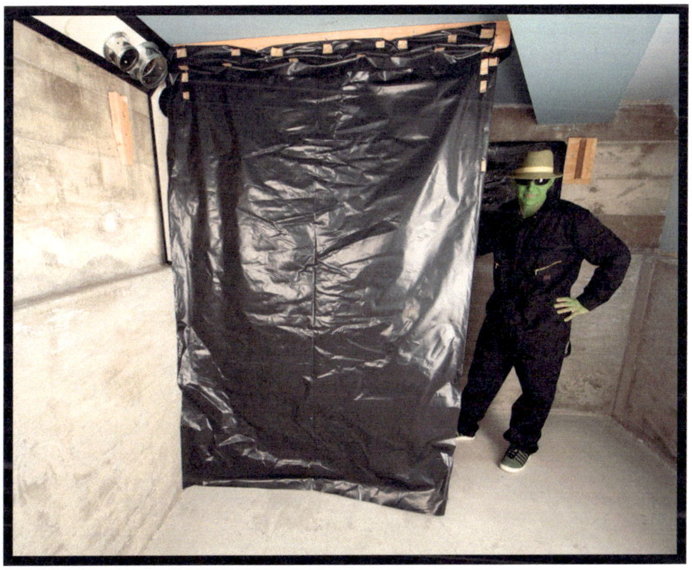

TOOLS: STAPLE GUN; STAPLES; VELCRO MATERALS: Black Plastic; Cardboard Cutouts.

The next step is to put up reflective white plastic all over both rooms.

I like to use this type of plastic because it reflects more light towards the plants which equals more bud, plus it makes for easy cleaning between harvests.

Now the room is clean
and
close to 100% reflective.

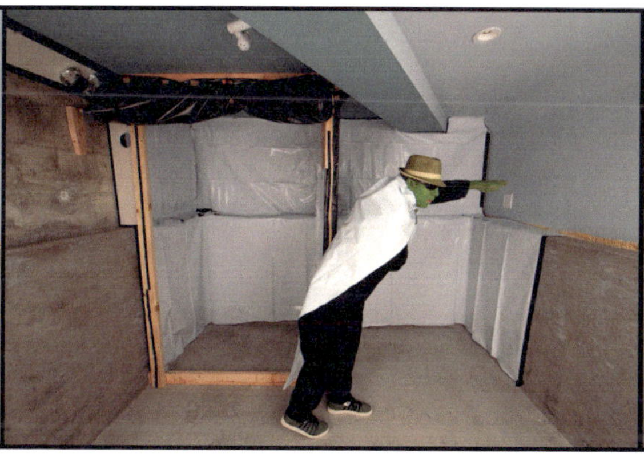

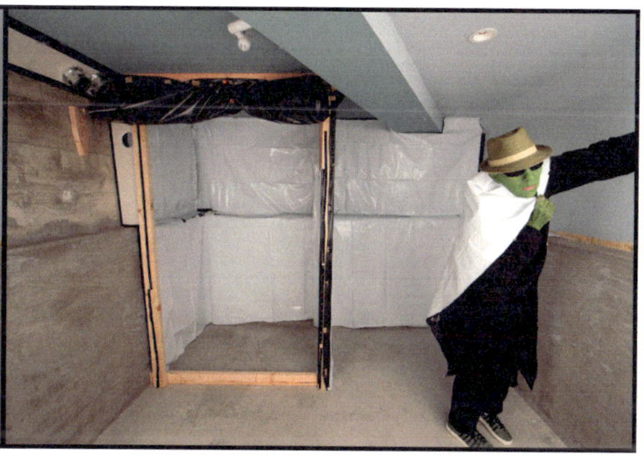

Lights up in the veg room!

These adjustable light hangers work fine with smaller lights like these.

It's time to hang up the T5 fluorescents in the vegetative grow room. These T5 grow lights are my favorite to use in the veg room because of the way my plants respond to them. Make sure to use the blue spectrum T5 'grow' bulbs for growing. This particular 2 foot by 2 foot T5 setup has eight, two foot long bulbs. These bulbs put out a nice cool blue light that the plants respond well to. I know this because of the way they rapidly start to grow, as well as the deep, solid green color in the leaves. Another reason to use T5 lights in the veg room is because it helps the plants keep tight internodal growth. Internodal is the space between growing branches. In this smaller type of growroom, it is important to keep the plants shorter and fatter rather than HUGE. The T5 light does a great job of keeping my garden bright and preventing the plants from stretching long and twiggy, as they can do when searching for light. I'm going to hang two sets of 2'x2' lights in this room. I'll start by placing the lights on the floor approximately in the area where the plants will be under. Then I screw a couple of hooks into some studs in the ceiling and gently raise the light off the floor, onto the hook, hung and done.

I always like to plug everything into the timers and test it as soon as I can. This way, if there is a problem, I can fix it before the plants arrive. Check, check, one, two... yeah, these lights are working fine.

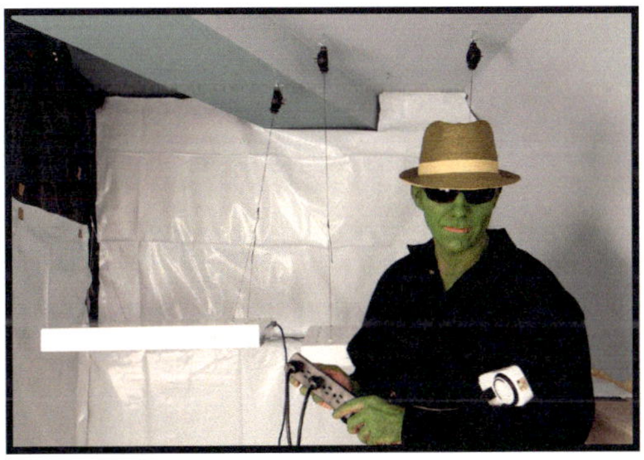

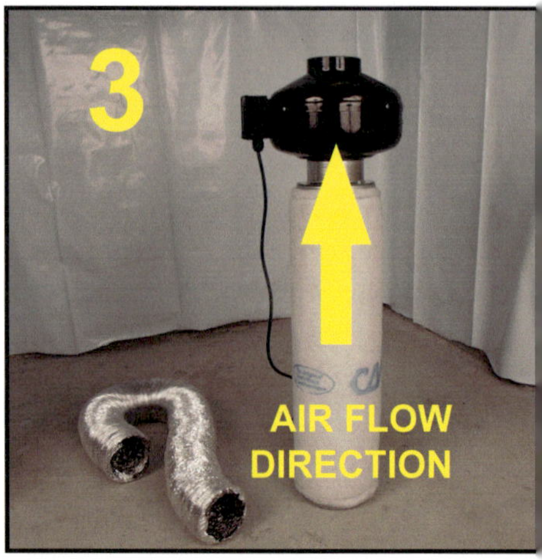

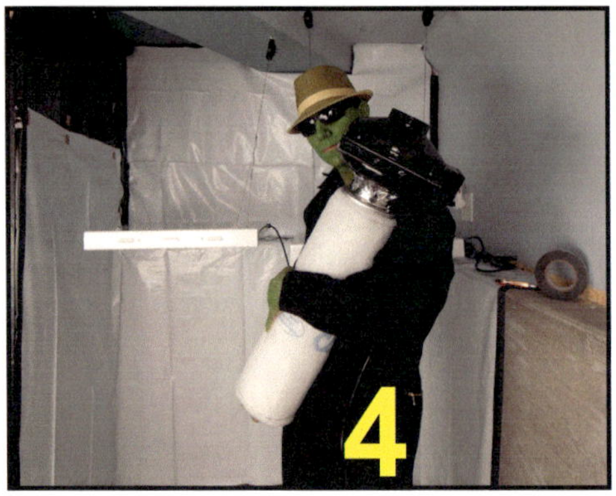

AIR FLOW.

1. 4 inch vent tube with 4 inch connector to mount onto the carbon filter.
2. This 4 inch vent fan is all that is needed to keep the fresh air coming into this space.
3. It all fits perfectly together.
4. Pick up the unit.
5. And hang it up!

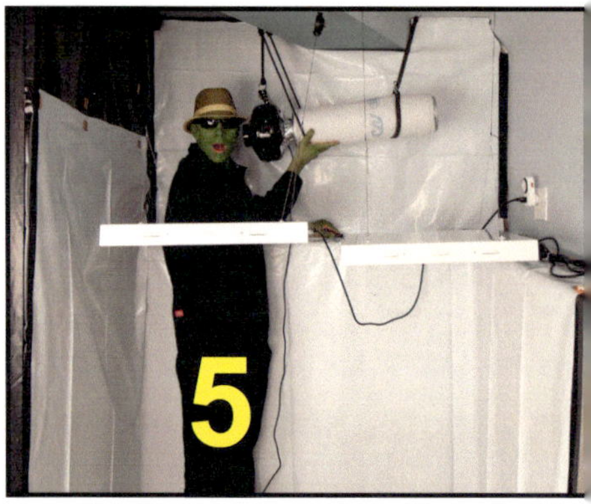

6 & 7. Locate the 4 inch hole cut out into the drywall and push the tube through it to the other side.

8. Secure with straps.

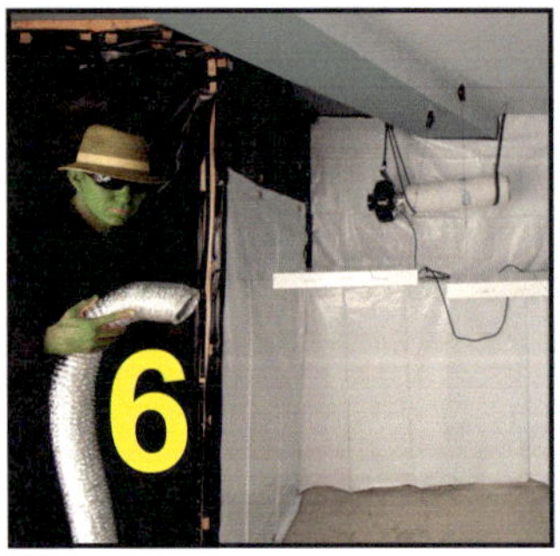

9, 10 & 11. I've connected one end of the 4 inch tube to the fan/carbon filter system and the other end to the vent outside.

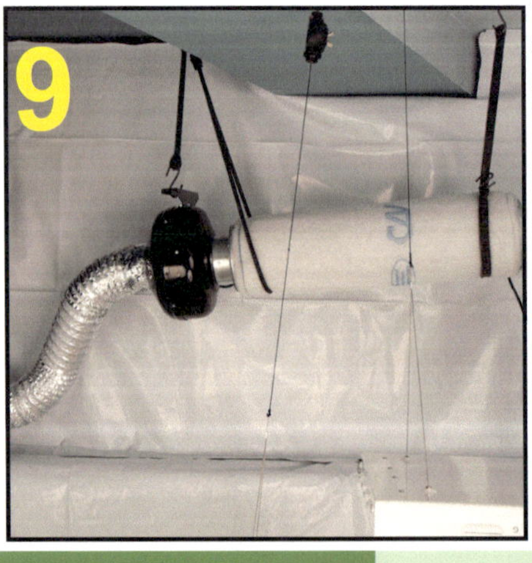

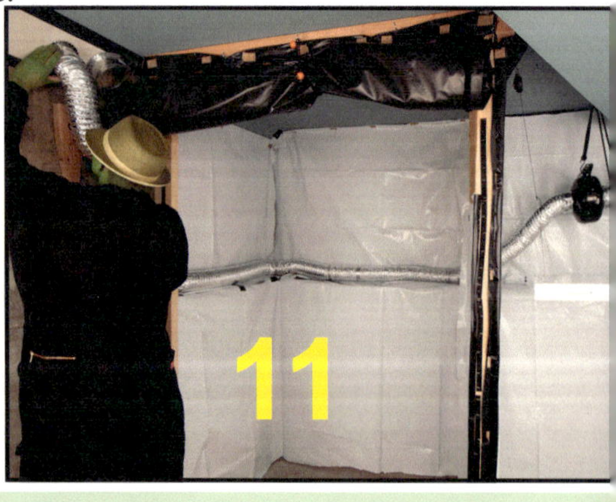

This is the completed ventilation system for the vegetative side of my growroom. I have it set-up so that the 4 inch vent fan will pull the air from the room through the carbon filter and blast it down the 4 inch vent tube to the outside. Even though this is not the smelly side of the grow, I still use the carbon filter to clean the air up because even a mild smell drifting to the wrong nose can ruin my secret garden. I can determine the square cubic footage of this spot by multiplying the height by the width by the depth. That's 8.5 feet x 4 feet x 5 feet or 170 cubic feet of air the vent fan has to clear out. It's good to know how to find the cubic footage of a room because it takes different sized vent fans to clear out different sized rooms. This 4 inch set-up is perfect for this small size room.

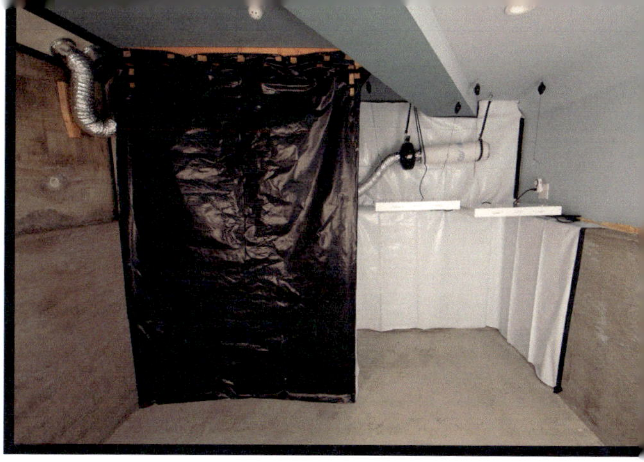

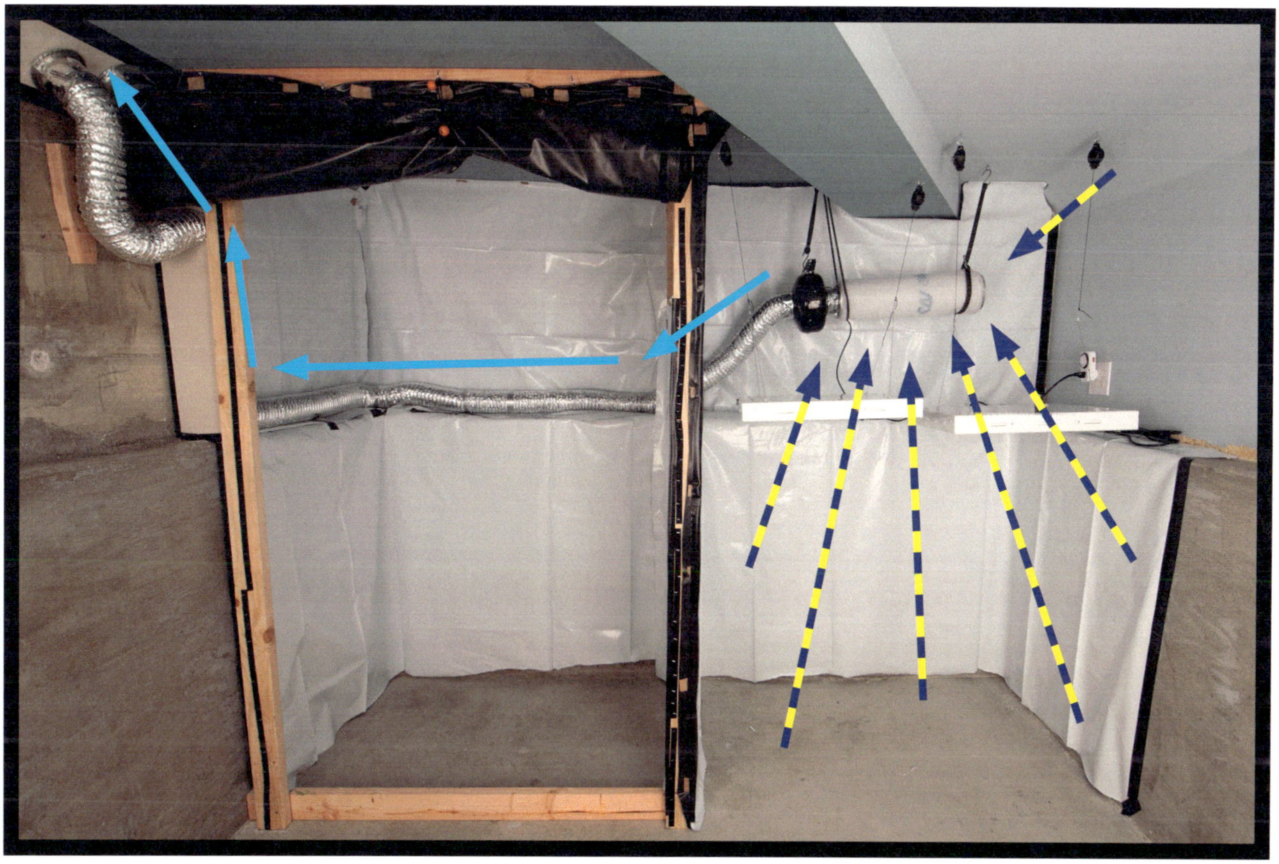

WATERING SYSTEM.

I've built myself a simple, yet sturdy, wood table to support the plastic tray loaded with plants.

I'm growing in these hard plastic, reflective-white grow trays made by Botanicare™. They last a long time and are easy to clean.

With the light system and air vent system set-up, I turn my attention to the watering systems. Not only do these systems deliver water and nutrients to the plants, they also act as the plant's home base. The first step for me was to build this small 2 foot by 2 foot table. The legs are about 10 inches tall. A notch is cut out so that the tray will fit onto it level. I drill a hole in the thick plastic tray and screw the nozzle into place. I connect a black plastic hose to the nozzle on one end and run it to a water collection tray placed underneath the table. Gravity will work to help drain off excess water from the plants and into the collection tray where I can empty it by hand. The set-up of the clone/ production plant tray is finished for now...

It is much easier to punch a perfect hole in thick plactic with a cone style, 'step' drill-bit.

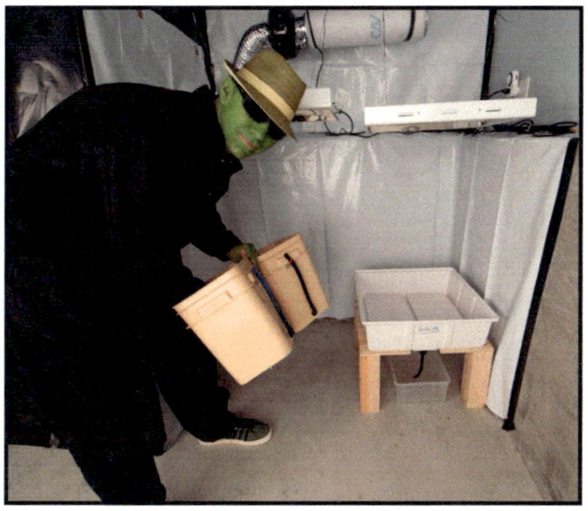

2 gallon buckets sit on the floor ready for the mother plants.

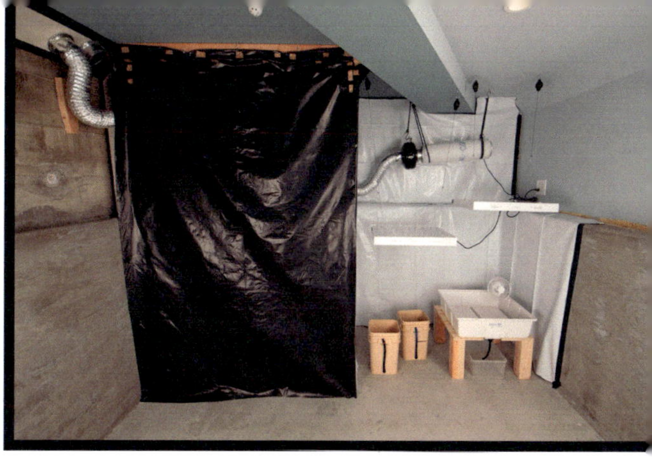

As for the Mother Plants; they are special because they do not go to flowering like the clone production plants do, but rather, stay in summer mode and parent ALL the clone production plants. The clone plants are a product of the mother plant. I grow my clones from the branches of the mother plant(s). Get it? I want the mother plant(s) to stay growing healthy in the vegetative room forever (or at least for a year). I keep the mother plants in these air-o-ponic Hydro-Farm™ buckets made by General Hydroponics ™. More on the inner workings of this bucket system when I bust them out and start growing with them later. For now, I place these buckets under the other T5 light and...

With the grow tray in place I'll hook-up this fan so that the room is flowing with a cool breeze.

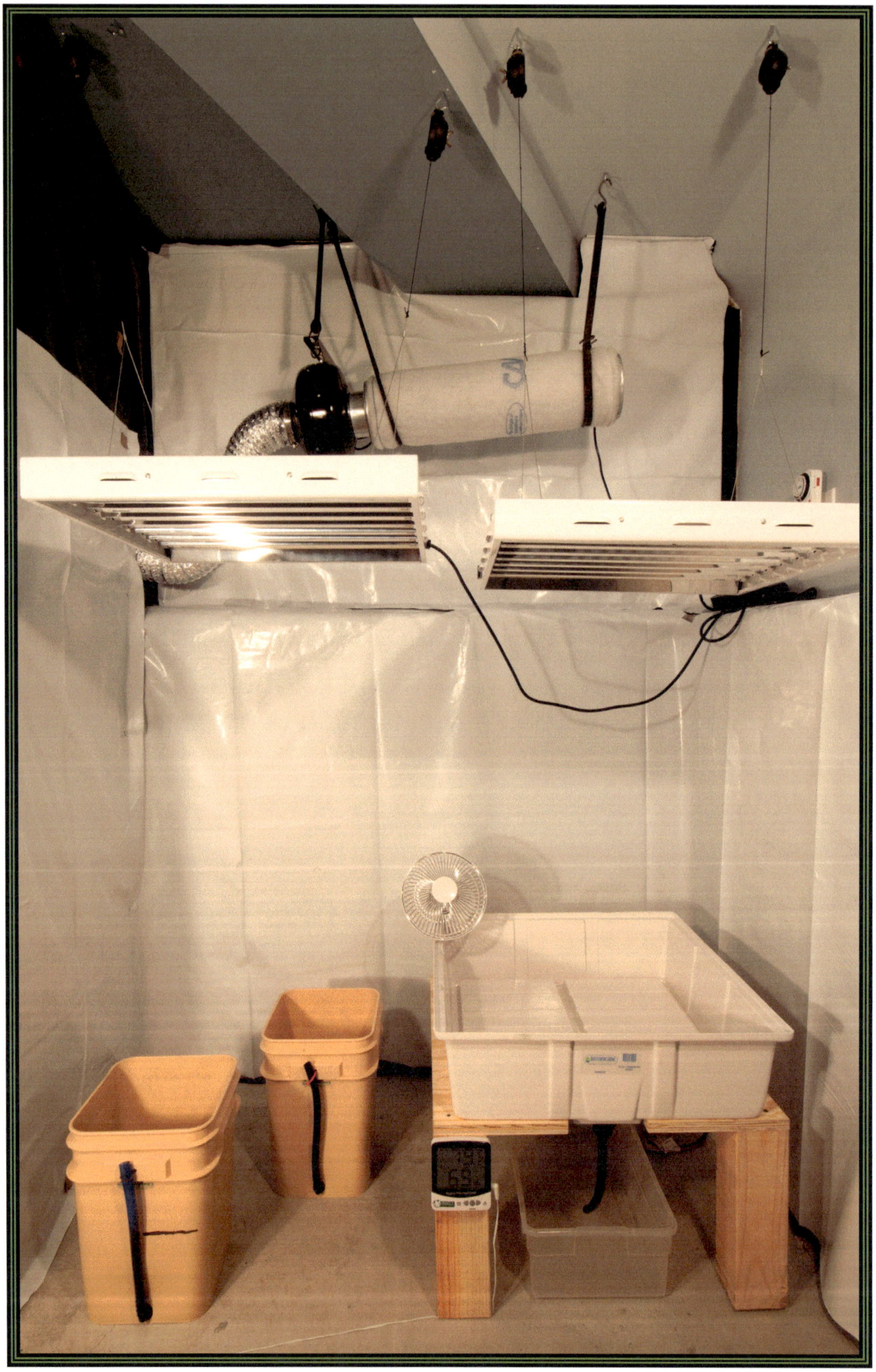

BAM!

The Vegetative Growroom Is Complete!

This is the new 'summer home' for the girls. I've paid special attention during the construction and have created a great environment to grow plants. I have introduced an air system that not only delivers fresh air to the plants, it also purifies the air of "green" odor before blasting it outside. The T5 fluorescent lighting system rocks. The blue spectrum of light they producc helps the plants keep the node growth tight. Tight nodes = tight buds. The hand watering system is all set.
Check-it... I've got a spot for the young plants and a place for a mother plant or two, which makes this side of the growroom a done deal!

LIGHT UP IN THE FLOWERING ROOM!

I make a shelf to place the light ballast.

Welcome to the flowering room. I'm going to use a 600 watt high pressure sodium lighting system. I'm opting to use a newer model electric ballast to power the system. Not only does this electric ballast run more quiet, it is 33% better on energy savings. Plus, this electric ballast has a much lower operating temperature than the old style of ballast, which is important for me, because I plan to keep the ballast in the same room with the plants. The closed reflective hood is a perfect size for this room and has easy access for bulb changes. The hood seals-up air tight with 6 inch vent holes on both ends for air cooling.

Yes, I am air-cooling the light because it makes a HUGE difference in my ability to control the room's environment. The HPS light puts off a lot of heat and will overheat the room without proper venting. The best way to solve this problem is to vent that hot air

This is my favorite kind of adjustible hook. They can support 50 pounds of weight and are easy to use.

I make sure not to touch the bulb with my fingers. Oils from the hand can cause premature damage to the bulbs.

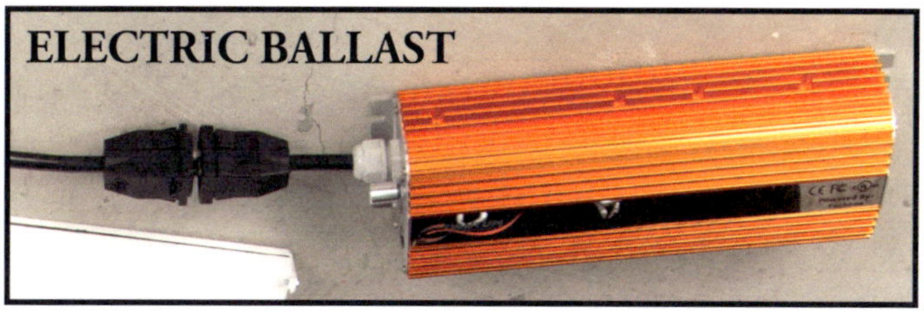

out of the room before harm can be done to the plants. That's why a nice, closed, refective hood is well worth the expense. The 600watt HPS bulb is a flowering, "red" spectrum bulb. I'll spend the extra money to get the "best" bulb, but the fact is some bulbs work better with certain ballasts. There is no real standard on what works best with what. For this room, I am using a Quantum™ ballast and a Hordalux bulb by GE™. Your local growshop will have all the details. I power it ON for a test... and it's looking good.

The deep orange glow produced by the HPS bulb.

AIR FLOW FLOWERING ROOM.

The budding room, inherently, has more challenges in regards to air movement. Firstly, it is going to get way more stinky in here! Secondly, with the big, stinky buds comes higher humidity and the potential for mold. The third challenge comes from the hot HPS light which needs to get vented to the outside. Looking at my inventory I have: 50cm carbon filter; 6 inch vent connector; 6 inch electric vent fan; 10 feet of 6 inch vent tubing; vent straps & duct tape. I'll screw the 6 inch vent connector to the top of the carbon filter and place the electric vent fan right on top of that, air-flow running upward. I lift the whole thing up and place it on the shelf in the corner next to the ballast. I make sure it's secure because I don't want it falling on the plants later. Now, all I have to do is connect these things together. I'll cut the 10 feet of vent tubing into two parts, about 5 feet each. I'll take a 5 foot piece and strap one end onto the exit vent fan and the other end to the light hood. Secure it. Now, I'll connect one end of the remaining 5 feet of vent tube to the second 6 inch hole on the light hood. The last connection is to the vent that leads outside.

I have to fit the vent hose through the 6 inch hole in the drywall and then make a hard right turn to the exit vent connection. Alright, let's go over this again... The air in the room is pulled through the carbon filter via the 6 inch vent fan and blasted towards the light hood. The air will have been cleaned of smell before it passes through the light hood cooling the bulb off. Then the hot air from the light is blown outside. Follow the blue arrows to see the air path to the outside. A direct path will give you more power, but the long way around with some twists and turns has the advantage

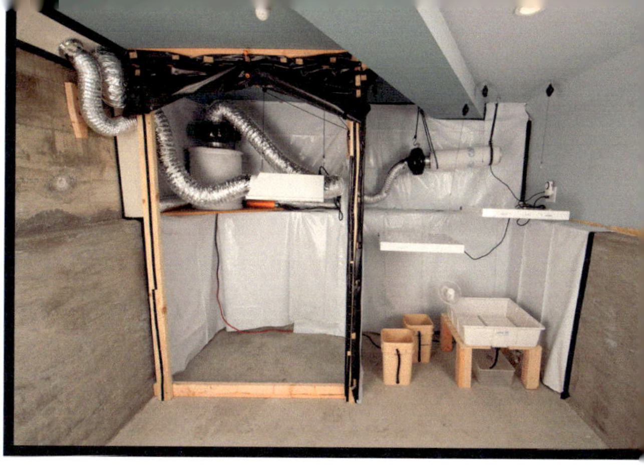

of keeping the fan noise level down at the vents end. Now that the air system and lighting system are all connected, I can turn my attention to the...

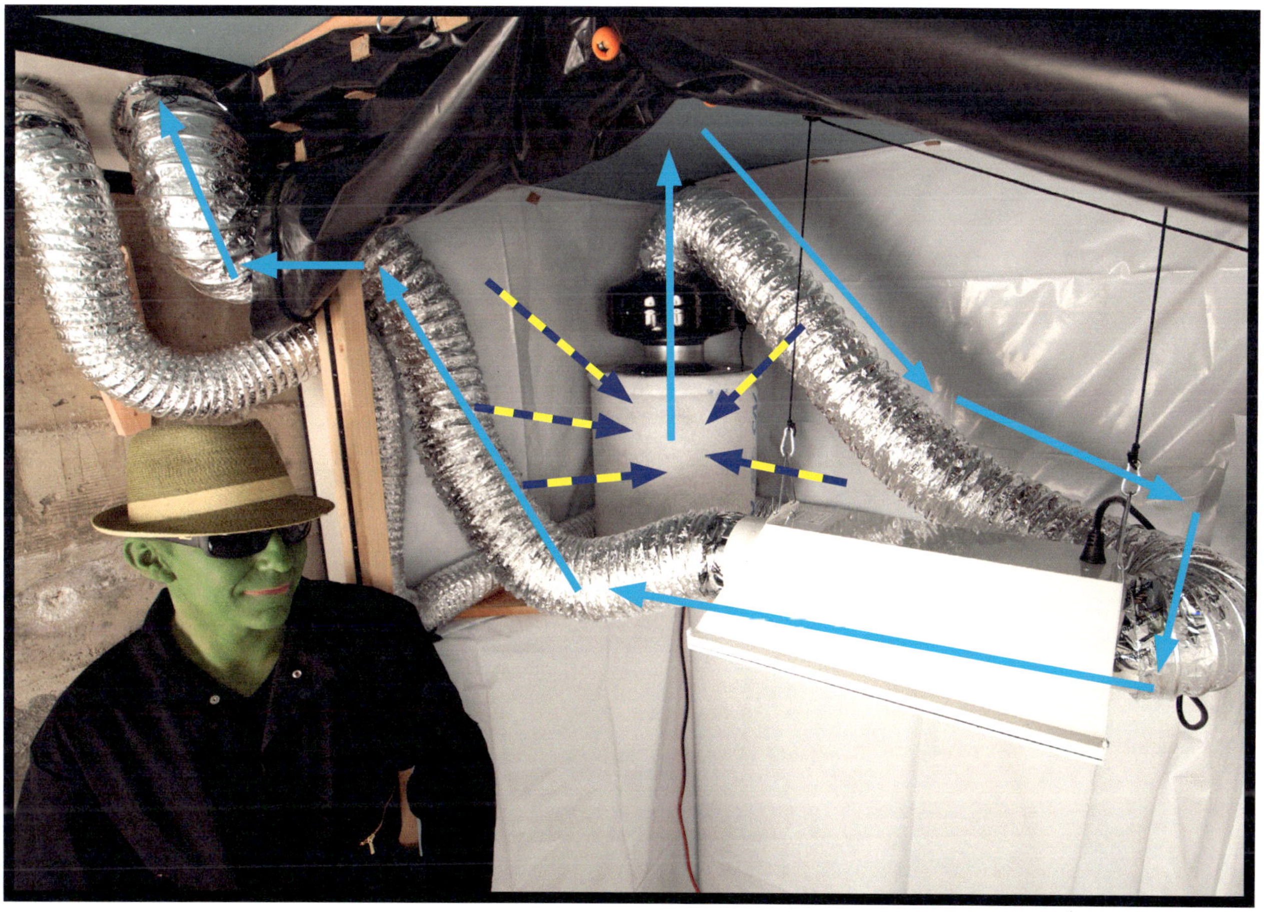

WATERING SYSTEM.

I am going to set-up the same type of water delivery system that I used on the veg-room side. I start by constructing a sturdy 2 foot by 2 foot table that the grow tray will sit on. I've cut a notch out of the table so that the tray will rest flush on it and drain properly. Now, a hole will need to be drilled into the tray and a nozzle will need to be screwed into that hole. It's important to make sure to cut the correct size hole for the nozzle I plan on using because these grow-trays ain't cheap sucka! I put it all together, connect the drain hose and put down the catch-all bucket to **finish the job**.

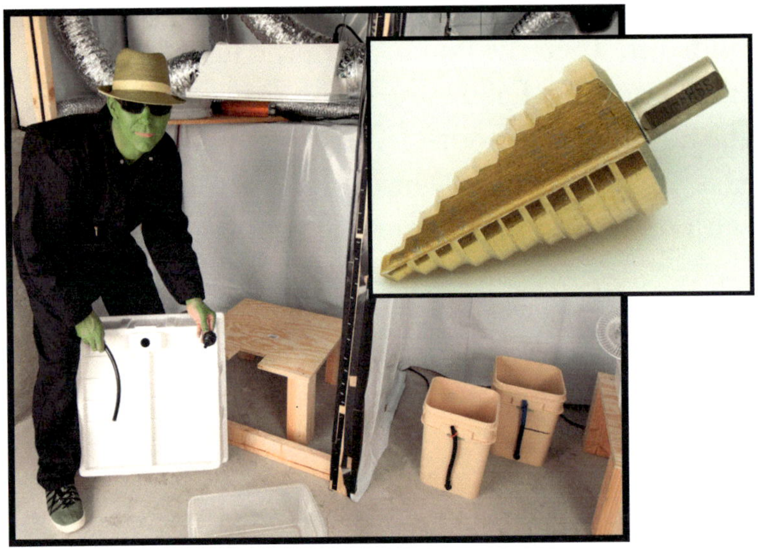

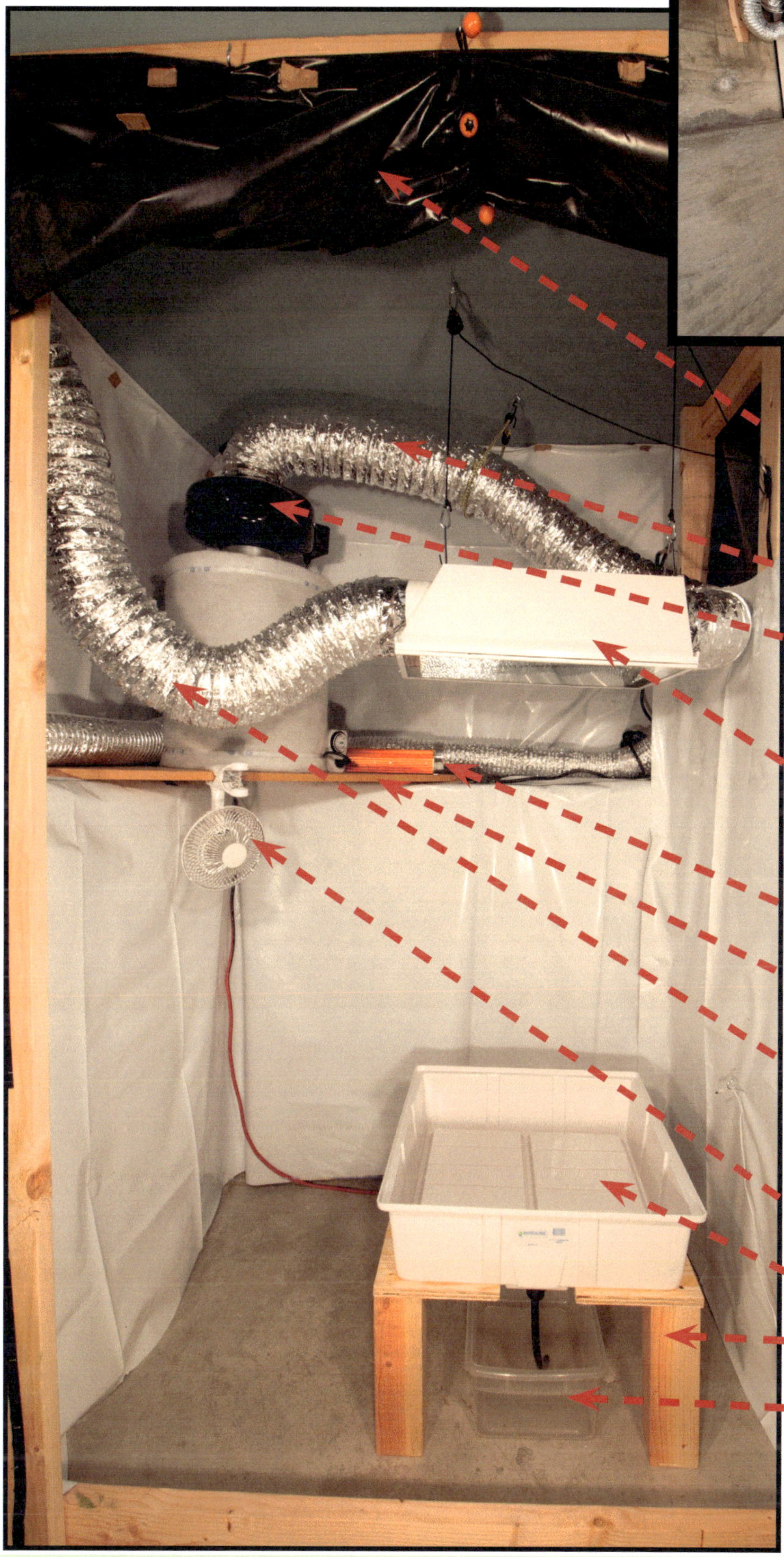

-BLACK PLASTIC ROLL DOWN DOOR.

-EXHAUST VENT HOSE.

-6 INCH VENT FAN WITH CARBON FILTER.

-CLOSED, REFLECTIVE LIGHT HOOD.

-ELECTRIC BALLAST.

-TIMER, POWERSTRIP.

-EXHAUST VENT HOSE CONNECTED TO EXIT.

-ROOM FAN.

-GROW TRAY.

-TABLE.

-OVERFLOW CATCHALL BUCKET / TRAY.

1. T5 fluorescents for veg room. 2. Small carbon filter. 3. Four inch vent fan. 4. Four inch exit vent for veg room. 5. 2'x2' grow trays. 6. Waste trays. 7. Hydro-buckets for mother plants. 8. Small room fan in veg room & one fan goes here for the bud room.

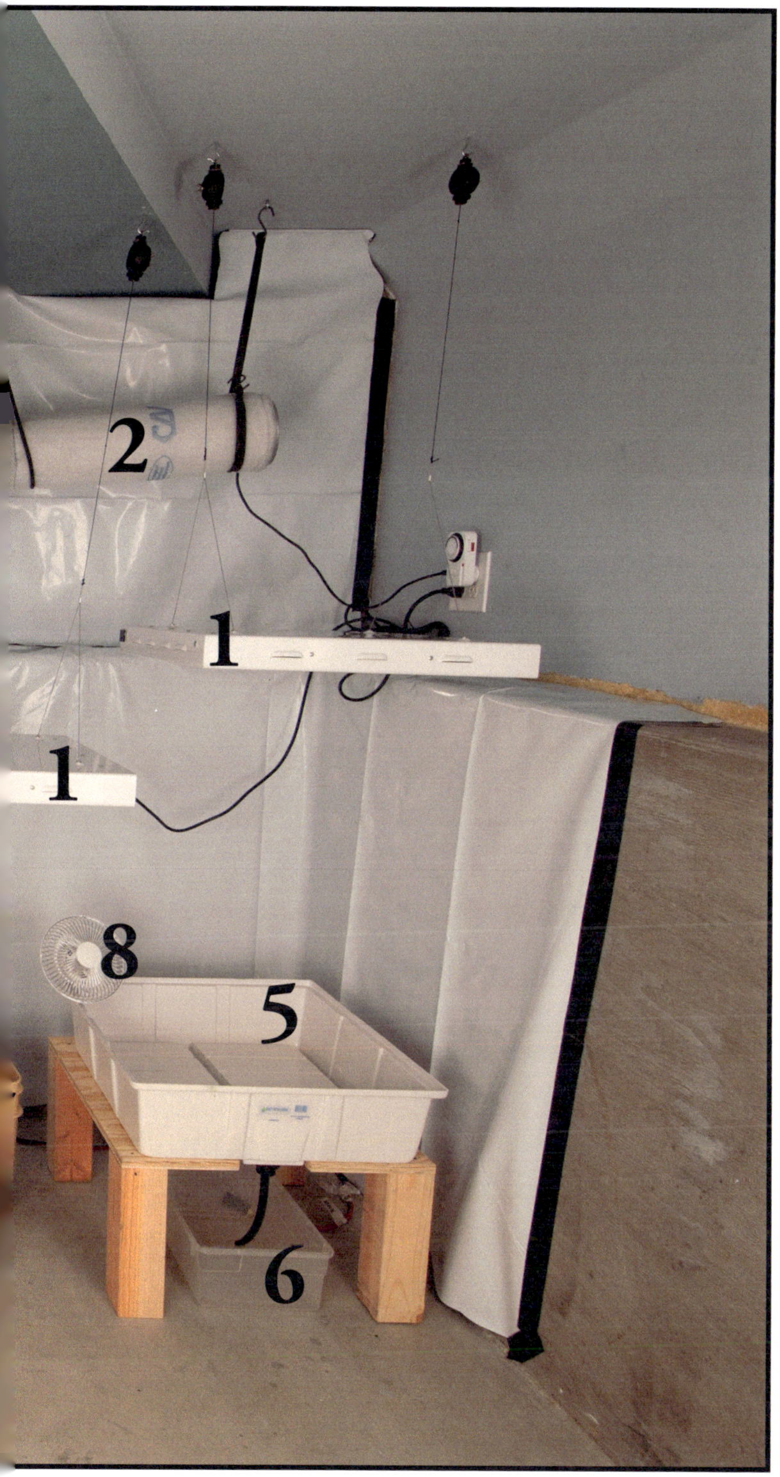

AND IT'S FINISHED.

The growroom construction is complete! It only seems like yesterday that I was contemplating on where to build my indoor grow and now I'm finished! Here, I've essentially built two small growrooms. The vegetative side, on the right, will be lit up 18 hours per day, imitating the long days of summer. The flowering room, on the left side of picture, has only 12 hours of light per day, mimicking the shorter days of late fall. The mother plant(s) will have a permanent home in the buckets and the clones are going to be grown in hydroblocks that make for easy transport from side to side.

9. Closed, reflective hood for bud room. 10. HPS electric ballast. 11. 50cm carbon filter. 12. Six inch vent fan. 13. Six inch air vent tube exiting hot air out.

With the *construction* of the growroom complete, I can focus 100% on *growing* cannabis. For first timers, the whole cannabis cultivating thing can be a bit mysterious.
SEE THE CHRONIC TIMELINE BELOW.
It's nice to have a little insight on what the journey to becoming a self sufficient cannabis farmer will look like.

First, I need to decide the kind of cannabis I want to cultivate. Will it be Indica, Sativa or a hybrid of both?

→ Acquire the proper cannabis genetics and start the garden from seed.

→ Germinate seeds, raise little seedlings and start growing young plants. (1 month)

→ Keep raising the young plants in a healthy manner as they mature and grow. (1 month)

→ Take clones from the plants started from seed 1st to determine their sex and 2nd for bud production. (1 month)

→ Flower (Bud) one clone plant from each seedling plant to see what sex (gender) they show. (1-2 wk)

→ Keep the female seedling plants! They are my new mother plants.

→ **RID the garden of male plants.**

This is the **growing cycle** I'm looking to be flowing into... When starting from seed the process will take at least 3 to 4 months just to get grown-up clones. From that point, it will then take 2 more months for flowers. That's up to 6 months before I'm harvesting bud! That's why some growers choose to get their clones from other respected growers for the first couple of flowering cycles while they raise their plants from seed (WARNING: this is the #1 way bad bugs enter and ruin gardens)... *Good clones are easy to find* but **great clones are very hard to find and are always treasured**. It may take longer to grow mother plants from seed but that's how to stay clean, independent and unique!

Take the female clones and grow them in the vegetative state for a couple more weeks, keeping them healthy so they can support large nugs. (1-2 weeks.)

I move the plants from the vegetative side to the flowering side where they will finish off their life cycle by growing magnificent flowers. (approx 2 months)

I'll take a fresh batch of clones from the mother plant(s), whom always remain in the veg room under 18 hours of light, about 1-2weeks into the flowering cycle so that the girls can grow ready for the next round of budding.

I continue to care for the newly rooted baby clones in the vegetative room for the next 6 weeks as the other plants finish budding in the flowering room. I'm working two rooms at once.

The plants are finished flowering and it's time for the harvest! This is actually a pretty busy time for most growers as we must trim up the bud and cure it properly, as well as, clean-up the growroom before transporting the next batch of plants in.

Now comes the fun part of deciding what strain(s) of cannabis to grow. This is a very important question because seed genetics are the #1 influencing factor in determining the potential quality of the flowers. Chronic bud always refers to top quality, grade A+, seed-less cannabis herb. It's important to understand that I can build a state-of-the-art growroom and give my plants the best water and nutrients the world has to offer, but I will still grow poor bud if I start out with lame genetics. I want to grow 'grade A' cannabis. This garden is for **Me**, so I'm growing flowers that get **me high** every time I medicate. I'm looking for a **powerful** plant, high in THC, to medicate and elevate. So, when shopping for genetics, I shop the more respected seed banks for their best and award winning strains. I've also been lucky receiving seeds from smaller companies and from hobbyists too. Getting excellent clones or seeds from friends is awesome, as some of the best herb comes from friends. Stoners and growers should know that there are basically two main types/strains of cannabis. Indica and sativa. Out of these two main types come thousands of different kinds of crossbreeds and hybrids. The amount of cannabis genetics available to the public today is a product of lifetimes of hard work by hardcore cannabis cultivating warriors. The trick now is to find the best seeds for my grow project from the many different choices of seed companies. Before I go on the search for strains, let me be clear on what I am looking to grow in this space: A mostly **indica dominant strain** that will remain **short and fat** while being covered in **very potent flowers** (15% to 20% THC levels). Plus, I'm looking for a plant that matures and ripens in **8 weeks**, give or take a few days. Some of the sativa strains and sativa dominant hybrids can get very large and take up to 12 weeks to finish! Sativas and certain hybrid mixes have many amazing qualities that I'm not looking for right now in this garden. The indica is what I want and need for this place.

I can go about my search for the perfect genetics in a couple of different ways.

1. I can fly overseas to Spain or Holland and visit one of the many seed brokers / distributors that will be more than happy to sell me genetics. I can even plan my trip during Cannabis Cup™ festival time and meet the seed companies in person.

2. Try Canada! Our Canadian brothers rock it! I'm sure to get lucky finding excellent seed brokers and seed genetic companies up in Vancouver, B.C. The only problem with Canada is that it's so nice I may never come back home.

3. The legal States of Colorado, Washington & California Co-ops, etc. As of January 2016, America has 23 states that have some sort of medical cannabis laws in effect. In California, for example, if you have a valid California I.D. and a recommendation from a doctor, you can shop any number of cannabis dispensaries. The problem is that most shops DO NOT SELL SEEDS. Shops claim they don't offer seeds because it breaks federal law but I believe shops don't offer seeds because they don't want you to grow your own!

I stay away from most "feminized" seeds because they sometimes carry a hermaphroditic genetic trait that even under minimal amounts of stress, may cause the female to produce male pollen. That is not cool, as the plant will start to screw itself until it produces its own seeds! The plant is then no longer seedless and not very good to smoke.

4. Friends. Friends can sometimes be the best place to find prime cannabis genetics. Friends often have clones to trade instead of seeds. This works fine because the friend usually has a history of working with that particluar plant and can give tips on how to keep the plant thriving and healthy. Sometimes there is an added bonus of being able to sample some bud before taking a clone plant home to grow. There is a risk of importing bad bugs into the growroom (see pages 101-102).

5. Cannabis Festivals and Expos. They can be a lot of FUN! They can also be a super lame waste of time! You must pick the right show in the correct location, as all cannabis shows are not equal and many do not sell seed genetics at all. You must choose wisely and GO! Bring cash.

6. Try the Internet. The world wide web has the largest selection of seeds. The only problem is that many companies will not 'officially' ship to the USA. So, buyer beware! And don't put yourself or your grow's location at risk.

VISIT www.igrowchronic.com for Mr. Green's seed reviews.

The cannabis plants I'm choosing to cultivate for this project are.....

Bubba Kush & Super Skunk Kush.

These breeds have everything I'm looking for in a plant:
- short overall height;
- quick finishing time in 56 days;
- fat, good looking, potent smelling, strong tasting buds; that are healthy, solid growers and easy to work with;
- each yielding between 1 & 1/2 and 2 ounces per plant;
- potent plants with a high THC content. Order from a trusted source and the chronic is guaranteed!

Bubba Kush> strong indica. Flowering time 8weeks/56 days. If grown in optimal conditions this plant can produce over 2 ounces of flowers per plant. The flower buds are very potent with a THC content of around 15%-20%.
Known to be Offspring of:
O.G. Kush x Bubble Gum.

Super Skunk Kush> strong indica. Flowering time 8weeks/56 days. When grown under optimal conditions this plant can produce up to 2 ounces of sticky bud. These buds swell-up with very potent resin crystals and up to 20% THC.
Her parents are known to be:
O.G. Kush x Super Skunk.

I also like choosing these two strains because, although they are a little different in taste and high, they are very similar in overall growing height and finishing time, which makes for easier farming.

Later on that day...
...at the local grow store
bloom
FLORALICIOUS
FloraBlend
FLORA NOVA

WTF?

Who knew nutrient shopping could be so confusing?

Shopping for nutrients for the first time can be a little overwhelming. There has been a rapid growth in the 'green' industry and with it has come a wave of new nutrient companies offering products that promise to grow "faster-bigger-better buds!" Many claims are marketing hype, but some are not. It is true that some nutrients are more superior than others. The fact is, all of these different product options can make your head spin with confusion. So what I'm going to do is try to break it down for you... Okay. When I go nutrient shopping for this project I want to make sure to bring some CASH! These **"natural, liquid, connoisseur quality nutrients"** are not cheap! With that said, I can not overstate the importance of excellent quality nutrients. I am confident in the knowledge that a plant's genetics contain that plant's inner potential and it is up to the grower to tap that potential and produce the most excellent herb possible. The first thing to do is to create a perfect environment, the second is to get great genetics and the third is to use top quality nutrients that work well with the life cycle of the plant. When shopping, I look for nutrients for both the vegetative and the flowering stages of growth. The different life cycles of the plant require different types of nutrients. I will be shopping for a **base grow** nutrient and a **base bloom** nutrient. When I use the words *base nutrient*, I am referring to the "main nutrient" but there are other nutrients that are needed, called micro-nutrients (additives). After choosing a base grow nutrient, I then select some additives that would be beneficial for vegetative growth, such as; *root boosters, carbohydrates, enzymes, and beneficial bacteria* for overall health and growth.

(Water + Base Nutrients + Micro-Nurtrients (Additives) + PH adjustment = completed nutrient batch)

I'll do the same for the flowering side by choosing a high quality, natural, base bloom nutrient and some high quality additives that help with overall health, flowering and resin production. I'm also picking up a flushing solution that I can use during the growing and flowering cycles, to rid the plants of toxic build-up. Without a flushing regimen, the plant's roots can build up nutrients to the point at which the plant's system **LOCKS-UP**. From there, the plant burns from the inside out as it quickly dies. So I need to purchase a good flushing agent, like Clearex™, along with everything else.

VEGETATIVE NUTRIENTS

(These are the nutrient products I used for this grow.)

1. PURE BLEND-PRO GROW™ By BOTANICARE® This is the base nutrient.
2. SWEET™ By BOTANICARE® The carbohydrates.
3. LIQUID KARMA™ By BOTANICARE® The sea kelp additive.
4. ROOTS EXCELURATOR® By HOUSE &GARDEN VAN DE ZWAAH™ An excellent root enhancer.
5. GREAT WHITE™ and/or **PIRANHA™** beneficial bacteria.
6. BLOOM, ROOTS™ By Agricultural Organics® (I alternated this product with Root Excelurator for the book but I now no longer use products from this company).
7. BLOOM, PRE™ By Agricultural Organics® (I used this product for this book but I no longer use products from this company).
8. CANNAZYM™ by CANNA. This is a plant enzyme product that helps the plants break down foods and digest.
9. pH-DOWN™ By ADVANCED NUTRIENTS® Must adjust pH just before feeding! This is a strong clean acid.

A main tool in growing superior chronic is the use of top-of-the-line nutrients. With years of trial and error I've developed my own recipe of nutrient mixes. When I first began growing cannabis through the filming of the "I Grow Chronic" movie, I used the pink, green & purple base nutrients made by General Hydroponics®. These nutrients worked very well for me for many years BUT I could not ignore how much better the quality of my buds were after switching over to a more natural mix of nutrients. The flowers just tasted better and the stone was clean and long lasting... Some growers believe that you cannot grow as big or as heavy of flowers with organic based natural nutrients as you can with chemical based "steroid" nutrients. The choice is yours but BEWARE! Some nutrients are TOXIC to humans! Nutrients that are labeled "for ornamental plants only" or "NOT FOR FOOD CROPS" are labeled so for a reason! I don't care if the grow store sales clerk promised a bigger crop or what, the fact is some "bad" nutrients can even put cancer causing elements into the bud! This is another great reason to be growing your own. This is no joke, growers want to stay away from PGR's (plant growth regulators), hormones and other dangerous nutrients. Companies can not always be trusted... for instance, I no longer use products from the "Bloom" line by Agricultural Organics (seen pictured below). My local grow store stopped carrying these products because when third party tested, they were found to have PGRs in them! I want to be safe and trust the nutrients I'm using. I'm big on quality over quantity, so I'm going to grow as natural as possible. I'm not growing 'organic' because I'm using a hydroponic medium and that, by definition, is not organic. But I'm going to try and grow as close to organic and healthy as I can. As it sits right now, this complete mix of nutrients will set me back about $500 US. That should last a couple of grow cycles. Whatever you use, make sure it's of high quality so that your buds will taste excellent and produce a clean and unique high. I love my flowers and find it hard to smoke anything else but my own bud.

(For updated nutrient product information visit www.igrowchronic.com)

FLOWERING NUTRIENTS

(These are the nutrient products I used for this grow.)

1. PURE BLEND-PRO BLOOM™ By BOTANICARE® This is my base nutrient for the first 3-4 weeks of flowering.

2. PURE BLEND-SOIL MIX-PRO BLOOM™ By BOTANICARE This is base nutrient for the last 2-3 weeks.

3. SWEET™ By BOTANICARE® The carbohydrates.

4. LIQUID KARMA™ By BOTANICARE® The Kelp mix.

5. BLOOM ULTRA™ By Agricultural Organics® Weeks 1-2. This additive helps keep the plants shorter. (I used this product for this book but I no longer use products from this company.)

6. BLOOM OOZE™ By Agricultural Organics® Weeks 5-6. (I used the product for this book but I no longer use products from this company).
I Use **HYDROPLEX™** By BOTANICARE® as a replacement.

7. GREAT WHITE™ and/or **PIRANHA™** beneficial bacteria.

8. BIG BUD™ By ADVANCED NUTRIENTS® Weeks 2-4.

9. OVERDRIVE™ By ADVANCED Nutrients® Weeks 5-7.

10. CANNAZYM™ by CANNA. This is a plant enzyme product that helps the plant break down foods and digest.

11. pH-DOWN™ By ADVANCED NUTRIENTS® 6.0 ph

Mixing & Measuring

Remember, lots of great farmers grow excellent, grade A, sinsemilla using their own style of nutrient mix or method. There are many ways to grow great herb. I have found a nutrient recipe that works for me and I'm happy to share it. It's exciting to see the nutrient companies putting out great products and my recipe / mix will continue to evolve. These are the exact products I used for this book. I am not sponsored, endorsed, nor do I have any financial interest in any of these nutrient companies. I do, however, like the way these products help my crop grow. Now that I have gone shopping and purchased these nutrient products, it is time to learn how to mix and measure them properly. The first rule to remember in mixing, is that you never want to mix the separate nutrients directly together because this can change the make-up of the nutrients, causing them to bind together and thus making them unavailable to the plants. I start the mix with a 5 gallon bucket of clean, fresh water. I measure out, then pour in the nutrients, one at a time, making sure to stir the water between nutrient injections. It's important to remember that in most recipes / mixes, in general, it's better to add less than to add too much, as you can always add more to a mix but once it is in, it can not be taken out. After adding and stirring the nutrients and additives, I take a PPM (Parts Per Million) measurement reading. If the PPM reads too low, I can add more nutrient. If it reads a little high, I can add more water to the mix to dilute. I would like to have the PPM at around 1000 for the vegetative stage and then have it slowly increase to 1500 PPM during the heavy flowering stages of mid to late bloom. Once I have the batch reading at my desired concentration of nutrients, I can turn my focus onto measuring and adjusting the pH. This is the last step in mixing the nutrient batch, just before feeding the plants, and it deals directly with the acid and alkaline levels of the batch. I need to get the pH level as close to 6.0-6.5 as possible. If the mix is reading high, such as 7.5, I will add the pH down, "acid," to the mix and stir it until the pH levels come down. If the mix is reading low, like 5.3, I would add pH up, "alkaline," to the mix. For me, the mix rarely reads low. I almost always have to add pH down. Everybody's water reads a little different. That's why it is important to test and adjust the mixed batch, before the feeding.

Also, try not to chase the pH. That is, mistakenly adding too much of one thing and then adding more of another thing to try and balance it out. For example: If I add too much pH down and it drops the pH to 4.0, I will not add pH up to try and get it back up to 6.0, but rather I will dump-out that batch and start anew.

Go to page 99 for the complete nutrient feeding schedule

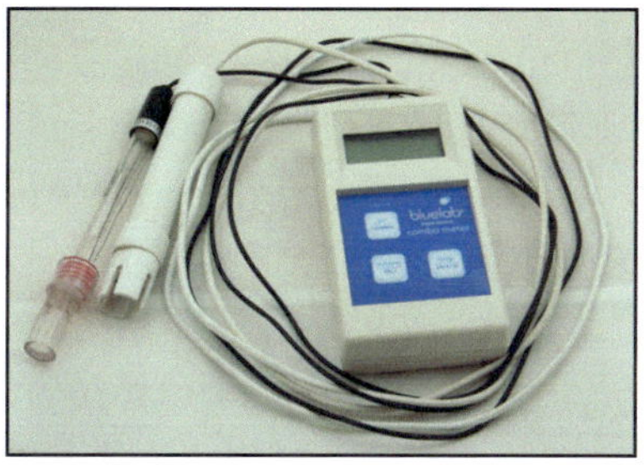

This is an all-in-one, pH, PPM & temperature electric meter reader device. I highly recommend using something like this although they can be costly. US $200-$300.

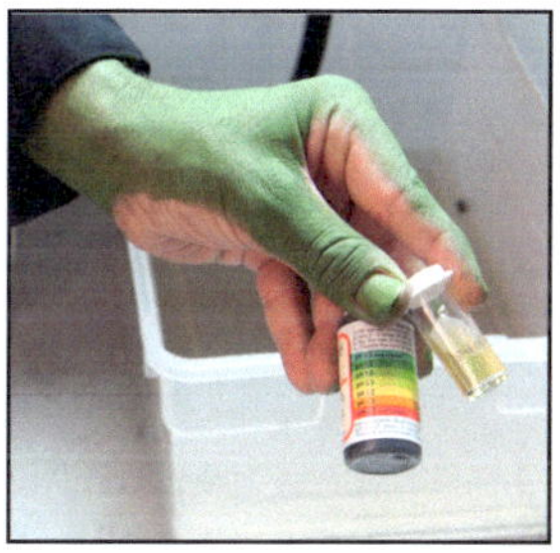

This is the manual method and it's a garden 'must have'. It takes more time to get a reading but it only costs $10 and it does not require batteries! I'll still want to purchase a PPM meter, and some sort of water thermometer. Approx. total cost US $100.

BASIC 5 GAL. NUTRIENT GROW MIX. ***

Start with 5 gallons of filtered water at 200ppm or less.
+ 100ml of base nutrient "Pro Grow"™
+ 50ml of "Sweet"™
+ 50ml of "Liquid Karma"™
+ 5ml of Roots Excelurator® or Roots,Bloom™
+ 25ml "Cannazym"™

BASIC 5 GAL. NUTRIENT BLOOM MIX. ***

Start with 5 gallons of filtered water at 200ppm or less.
+ 120ml of base nutrient "Pro Bloom"™
+ 100ml of "Sweet"™
+ 50ml of "Liquid Karma"™
+ 50ml "Ultra or Ooze"™
+ 50ml "Big Bud or Overdrive"™
+ 25ml "Cannazym"™

*Water purity levels vary from place to place and from time to time, even in the same place. That's why it's important to do your own PPM testing.
**Nutrient companies try to produce their products as consistent as possible but there can often be slight differences in their mixes or a mix 'change' altogether. That's why it's important to test the nutrient mix by taking PPM & pH readings.

THE WEEKLY NUTRIENT FLUSH

There can be too much of a good thing and it can be deadly to the crop. Nutrients. The same ingredient that supports life and growth can also deliver death. It's called nutrient lock-up and it happens when nutrient levels exceed the healthy limit and enter into the toxic zone. In order to prevent this from happening, I take the pro-active approach of flushing my plants root system every week. Not only does this wash away any unwanted build-up, it serves to freshen the plants up and seems to give them a lot more energy.

TIP: I foliar spray the plants down with pure water before pouring "Clearex"™ over the root cubes. I'll use 15ml of flushing agent per gallon of water as to really clean the roots out good.

BENEFICIAL BACTERIA & TRICHODERMA ADVICE.

After the plants have been flushed out, I feed them beneficial bacteria & trichoderma. I do not add them to the nutrient batch because the nutrients will burn-up & kill most of them. So, I mix them separately in pH6.0 water and feed the plants by hand so that the beneficials have a chance to establish themselves in the root systems. I'll wait at least 24 hours before I start feeding the plants nutrients. This is something I will do every week during the vegetive grow cycle and at weeks 1 and 4 during the flowering cycle.

GERMINATION DAY.

All the work and planning has been leading up to this day, germination day. I've got the seeds and now the goal is to grow at least one, maybe two, great female plants. I'll start with germinating 10 of these seeds. Once they sprout I'll care for these baby plants in the vegetative grow room. Soon the baby plants will start growing into medium sized plants. The plants will be needing more attention with more frequent waterings. After the first couple of months of growth I want to be able to determine the sex of the plants. I only want juicy, fruity, female plants in this garden. Remember, male plants produce the pollen that can impregnate the female flower bud. Once the humping has taken place the female plant stops producing THC covered flowers and starts putting her energy into seed production. The way to figure out the sex of the plants grown from seed is by taking a clone (cutting) from each plant and then budding that clone out in the flowering room. It takes about 10 days before the sex of the plant can be seen and discerned. Once I know the sex of the clone, then I know the sex of its' corresponding mother seed plant. (This is why I LABEL all the plants... to avoid confusion). After I determine what plants are the MALE plants, I can rid them from the garden. If I have too many females to choose from, I will rid the garden of the weak ones. I will be left with only one or two healthy females which will be my Mother Plants. I'll take all my future clones for flower production from these Mother Plants, that will be growing in the endless summer of the vegetative grow room. Now that you know the plan, lets get back to the seeds... I only want to start with 10, so I'll pick 5 of the best looking Bubba Kush seeds and 5 of the best looking Super Kush seeds. Good looking seeds are hard and solid with a nice dark color to them. Bad seeds "pop" dead when squeezed and look pale and weak. I get two small cups and label them, one bubba kush and one super skunk kush, as not to mix the seeds up. I fill the cups halfway with distilled water and plop the seeds in for a swim. Filtered water is okay to use in a pinch but it's important to remember that seeds, as well as the baby seedling plants, do not need nutrients. All the nutrients needed for the newborn is self provided by the seed and is consumed by the baby seedling during the first ten days of growth. After that period of time I'll give the tiny plants a weak nutrient solution mix, 20% strength (200-300ppm grow-mix). Once all the seeds are in the water, they should sink to the bottom of the cups. At this point, I'll take the labeled cups with the soaking kush seeds and place them in a dark, quiet, safe spot until they hatch. I'll start checking the seeds after 24 to 48 hours for signs of life. Some seeds visibly crack open after only a few hours and other seeds take a couple of days, even a week to open up. Once I see the tap root exposing itself through the crack in a seed I gently take

hold of the seed with clean tweezers and place it into a pre-soaked, 1 inch grow cube. I'm mindful to put the seed into the hole, in the cube, with the tap root pointing downward. I then place the whole tray back into a dark place and keep checking to see when the seeds start to push themselves upward in search of light. **This part of the process can take a few days.**

GERMINATION OVERVIEW.

Step 1) Plop the seeds into a cup of distilled water. Put the cup of seeds in a dark place and leave them alone for 24-48 hours.

Step 2) Start checking the seeds to see if they have cracked open and started to come to life. This process can take between one and several days.

Step 3) Once the seeds have absorbed water and have opened-up, gently gather them and place them, tap roots down, into some wet, 1 inch starter cubes.

Step 4) After the cubes have been seeded, it is important to place them back into a dark area and wait for the sprouts to shoot up in search for light.

ROOT INFORMATION.

The overall health of the plant depends on the health of the plant's root system. Mastering the root system is one of the most challenging aspects of growing. The two most important things to focus on are pH and oxygen. I make sure to adjust the pH of the nutrient water before I feed the plants. The roots grow into a rockwool cube and can get lots of oxygen in this fiber type environment. It is important for the roots to avoid extreme temperatures! Also, never let the cubes or roots dry out. Also, do not keep the cubes too wet and soggy thus prone to mold! Think, **"wet to moist, then wet again"**. Here's a list of things a beginning grower can do to keep the roots in top shape.

1. A weekly flushing of the root system clears the plants of nutrient build-up.
2. Try using a root additive like ROOT EXCELURATOR™.
3. Feed with water at 68 Degrees temperature.
4. Try beneficial microbes, GREAT WHITE™ & PIRANHA™ to enhance the root environment.

DAY 6

This is the first day under the T5 light. The light should be at least 10 inches from the tops.

Nice! It has been six days since my baby seeds started to crack open and today they have started to sprout upwards in search of light. Six out of the ten seeds I started with have sprouted successfully. Up until now the seeds have been kept in a dark warm place. Once they start to sprout, I put them in the vegetative growroom under 18 hours of light. I try to help the little sprouts out when I can by supporting them (paper clip), shedding the shells and keeping the cubes moist. Remember, no need to feed nutrients to the baby sprouts. Just water them with filtered water. The sprouts are very delicate right now that's why you don't want to put the light too close or touch them too much. Notice the babies at the bottom are having a hard time. It seems that sometimes, no matter how hard we try, they just don't make it. Such is life.

Don't let the cubes dry out, nor have the lights too close to the tops!!

DAY 8

Time to transport the sprounts into bigger blocks.

DAY 12

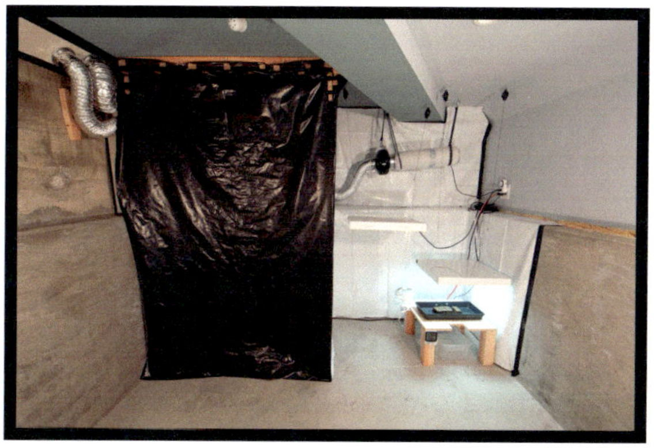

During the second week of growth, I can see the tap roots growing through the 1" starter cubes. That is a good sign, telling me the root system is developing. The roots will need much more room to grow so I'll transport them into bigger cubes. But wait, before I just plop them into the 3 inch x 4 inch grodan® rockwool cubes, I should 'set' (prep) the big cubes first.

Yes, there seems to be a trick to everything. Getting the Grodan® Delta™ cubes ready for the plant entails soaking the cubes in 5.5pH water for 30 minutes in order to 'set' the cubes. Once the 3 inch x 4 inch cubes are set, I simply plop the smaller starter cube into the pre-cut hole in the bigger cube and leave it alone. Once this part of the job is complete, I'll not feed the plants for 3 or 4 days while the roots rapidly grow.

This one didn't make it.

DAY 20

The sprouts are finishing up their third week of growth. They are between 3 inches & 4 inches tall and are putting out multiple sets of leaves. Now is a good time to start giving the sprouts a mild dose of nutrients during waterings. At this age, the babies are still tender and will burn-up if given an adult dose of nutrients. Either make a weak batch or dilute a regular mix with water to lower the parts per million of nutrients to an overall level of around 300-400PPM.

DAY 22

Once I start to give the sprouts a mild dose of nutrients, they grow bigger almost over night. I can notice the leaves getting darker, which is a good sign that the plants are uptaking the nitrogen out of the nutrient batch.

DAY 24

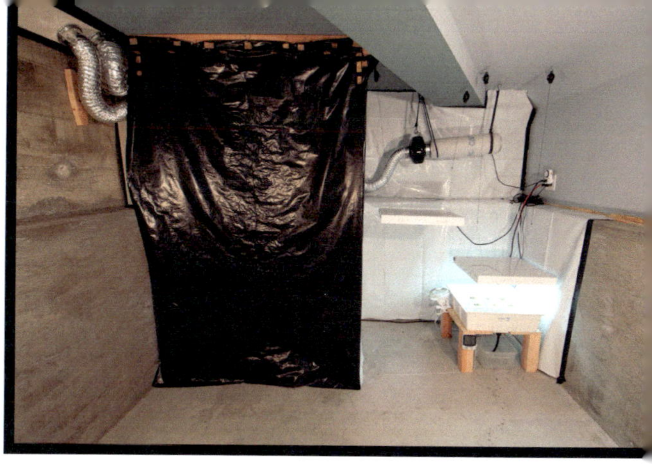

Out of the 10 seeds I started with, I now have 5 healthy little plants. All I'm looking for is one or two good mother plants, so the odds are still in my favor.

These baby plants are under 18 hours of light per day. The T5 fluorescents are now 6 inches above the tops.

**Notice the white reflective plastic on the tops of the cubes. These help prevent green algae from growing on the tops of the cubes.*

DAY 27

Concluding it's first month of life, this baby sprout stands 6 inches high and has developed 5 sets of leaves. During the first part of the next month, I'll be feeding the plants with 3/4 strength, 750PPM, nutrient mix every 2-3 days. I will soak the cubes and then let them slowly dry. When the cubes are still moist to the touch and about half of their max wet weight, I'll feed them again. I never let a cube completely dry out! If not watered, the plant will eventually drink everything within the cube and become very light to the touch. Soon after that, the plant will start to wilt, then die. Prevent this tragic event by watering the cubes "from wet to moist and then wet again."

DAY 33

The young plants are now into the second month of life and are starting to grow faster. At this point of the grow, I can tell the plants need more space for their roots to expand. But before I transport any of the five

potential mother plants into big buckets, where their roots can grow crazy healthy, I want to determine their genders. The problem is that the plants are not quite ready to be cloned, but their roots have outgrown the 4 inch grow cubes! The plants need a little more growth before I'll clone them. In the meantime, they will be needing more attention. The plants will need more frequent waterings and a little stronger mix of nutrients (800PPM+). It is not healthy to have the root system bound-up, but this is temporary, just until I find out which plants are the females and select the mother plants.

DAY 39

By day 39, the Bubba & Super Skunk Kush plants have grown over 10 inches. The root systems of these plants are being pushed to the limit and would really benefit from having more space to grow, but as long as I continue to water them and maintain this nice environment the plants will continue to grow and should be ready to clone in a few days.

Hmmm, I do believe these plants are ready to be cloned for the first time. I'll be taking one cutting from each plant, then budding that cut clone to SEE what type of flower it produces. Once I've determined which ones are the female plants, I will put them into their permanent homes, in the Hydro-bucket™ grow system. Although the plants are temporarily root bound, they will bounce back after a few days in the new system. Cannabis is a tough plant.

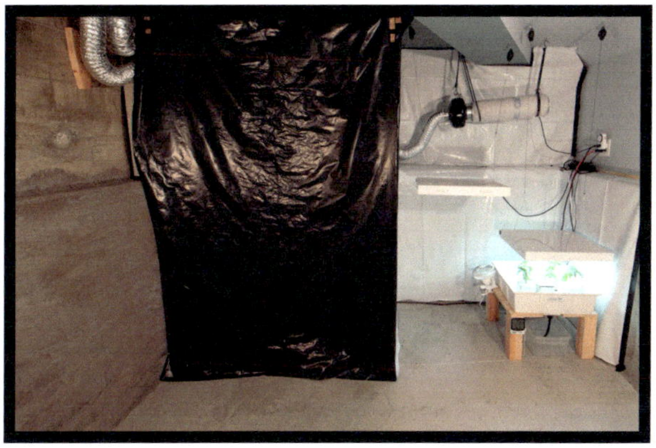

DAY 43

DAY 45. Sexing the plants! Gender identification clones are the first clones removed from the plant. I'm looking for one (or two) awesome female plants to call Mother. I'm not interested in growing males, so I will take them and remove them from the garden. The first thing I need to do on my quest is to take cuttings of these plants that I started from seed in order to determine the sex / gender of these plants. **(CLONING For gender identification).** Taking clones for gender identification is not different than taking clones for flower production except that there is little time spent in vegetative growth as I put them directly into the flowering room. It has been 45 days and the plants have all grown between 12 & 18 inches. That is plenty of healthy growth to clone them without causing harm. I try to never cut

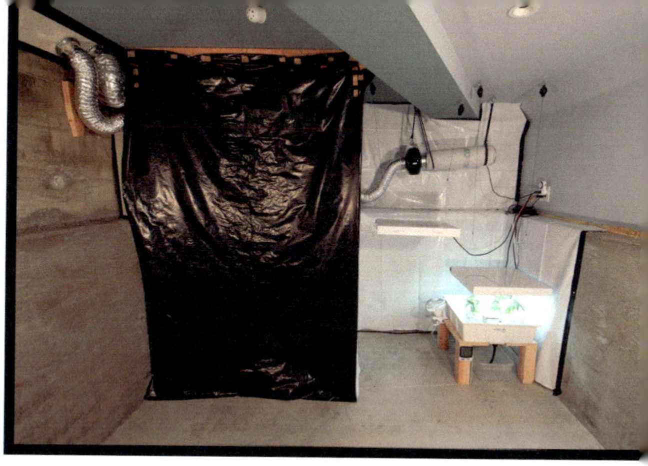

more than 1/3 of the foliage off of a plant because that might thrust them into a shock, from which they may not come back. Before I get chopping, I've made sure to label and tag all of the plants and all of the clone cubes. It is imperative to label all of the plants and keep order or else risk confusing one plant for another, thus voiding the whole sexing observation. (1) I get all of my cloning gear ready, including: wetting my 1" Oasis Cubes™ with water; cleaning my sharp scissors with rubbing alcohol; filling the spray bottle with water to mist the rootless plants; and opening up the jar of rooting agent (Clonex™). Now is the time to make the first clone. (2) I'll go about 3 nodes down from the top of the plant and remove the branches. (3) Next, I cut the clone from the plant, just under the previous cuts and at a 45^{0}degree angle, before (4) placing it into the Clonex™ gel. After dipping the cut end of the clone into the gel, making sure to saturate the cut end, I plop it into the hole in the cube. (5) I go about taking clones of all the plants. (6) I put them inside a dome tent. The dome tent is necessary to keep the humidity very high because the plants can only survive by drinking water through their leaves. No tent, no humidity, no survivers. I put them on the vegetative side for a few days until the shock wears off and they stabilize.

DAY 55. It's been 10 days and I've had great success with my clones throwing roots. Before long I'll know the gender of all the plants that I started from seed. I gather up the 5 rooted clones and place them on a one inch coconut fiber grow mat. I'll have to cut the coco mat to fit in this small grow tray. Putting down this coco fiber mat gives the roots somewhere to grow during the next couple of weeks, until I can discern what type of flower they are growing. I place the entire tray into the flowering room and set the light for 12 hours per day. The idea is to SEE what kind of flower each clone shows, so that I will KNOW the gender of its corresponding seed plant. After the plants are set up in the room, I feed them with a nutrient mix of 1000PPM for budding flowers. Within two weeks I should have my answers...

DAY 65. Let's start by examining the plants, searching for pre-flower calyx (pods) along the intersections of the branches and the main stem, the nodes. I'm looking for a pod with one or two white hairs (stigmas) growing out of it. If a plant has a calyx with a stigma(s), then it is a female plant. Pictured below are two female preflowers.

GENDER ID

This is what I have been looking for! Notice the white stigmas growing out of the calyx, indicating that these are female plants!

On this page, I have pictures of the male flowers produced by three different plants. Male flowers start off as tiny little closed flower pods in their early stages and then they bloom and release pollen into the air in search of female flowers. Now that I can see three of the five plants that I started from seed are males, I can remove them from the garden. That gives me two female plants to call Mother; Miss Bubba Kush and Mama Super Skunk.

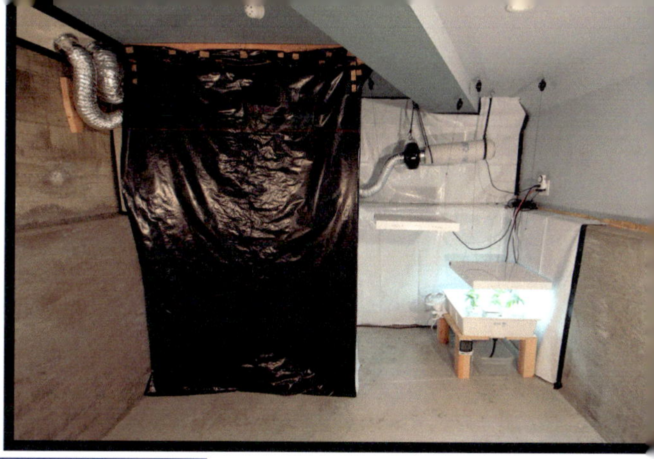

Notice how male pollen flowers form in clusters of pods off the top and off the nodes of the plant. Also, notice how much more developed the male buds are than the female buds. This is because the males mature faster, so that they can pollinate as soon as the female flowers start to bloom open. I will have long removed the males from the garden before that happens, thus insuring a seedless crop.

Mother Plants.

Day 69. I'm going to transport the two female plants into their permanent buckets and remove the males from the garden. I like these air-o-ponic buckets because they give the plants plenty of space and support for the root system. I think this style of bucket makes for a great long term home for a mother plant. They are simple to work with. I can flush out the plants and change the nutrient mix with ease.

I start by washing down the equipment and flushing out the rocks/pellets with water to remove any residue and excess dirt. I assemble the bucket unit and fill the top part up with some Dyna-Rok II™ (silica rock growing meduim) or Hydroton™ (Orchid potting pellets) or a mixture of both like I did for these plants.

I pick up the female Bubba Kush plant and remove the plastic wrap to expose the root system. The best time to transport the plants is as soon as the roots start to grow out from the bottom of the cubes. That was back a few weeks ago for these plants. They are now a bit root bound. The roots will start getting healthy after a few days in the buckets. I'll place each plant into a bucket, one at a time and fill-up the sides with rocks. The plant grows in the top part (red) of the bucket and the water/nutrient mix sits in the bottom (tan) bucket.

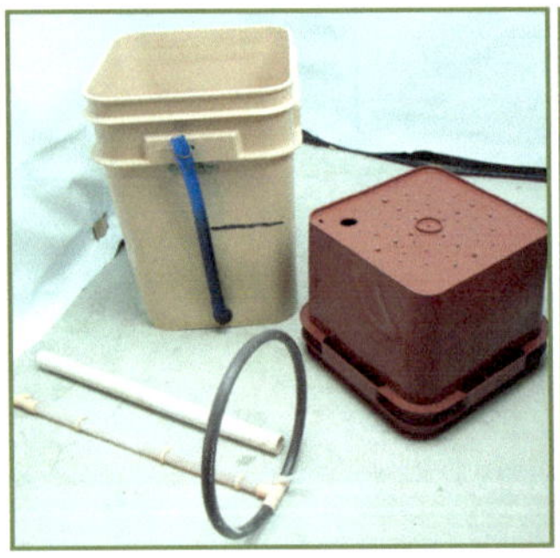

*NOTE: Silica rock seems to be getting hard to find in the stores these days, as a substitute try soaking any rock medium in a **mild** mix of liquid silica nutrient product for similiar results.*

Remove the protective plastic wrap and place the root block in the bucket under a drip hole to insure the plant gets water while the roots grow huge.

These buckets can be used and re-used for flowering the plants too!!!! The air-o-ponic design makes for excellent flower development! The downside is the initial cost, BUT in the long run it is actually cheaper than buying rockwool cubes every couple of months. I wanted to show two different growing methods in this book so I grew my mother plants in these air-o-ponic Hydrobuckets™and I grew my production plants (clones) in rockwool, Grodan™ cubes. You can make the right choice for your garden.

1000ppm Nutrient mix 1Gal
15-20ml Pro Grow™
10ml Karma™
5ml Sweet™
5ml Cannazym™
1ml Root Excelurator™
Adjust pH to 6.0 - 6.5

An air pump works to aerate the water that it delivers to the plant via an inner tube. The roots grow like crazy good in this system and should be super healthy within a few weeks! A mother plant is a perfect, female plant that is kept under a summer like environment for the sole purpose of having cuttings, called "clones", taken from her for flower production. The clones are an exact genetic copy of the original mother plant. They are the same plant living separate lives. There are huge advantages of having a mother plant(s), instead of starting from seed every time. Such as the huge savings in harvest turnover times and in overall work load. It saves at least one month not having to start from seed every time I need bud. This is a big reason I have a mother plant. I just demonstrated the process of growing plants from seed. It can take 3 months, whereas I can root a clone in 10 days and have it ready for flowering in less than one month. After two months of working with the seed plants, I didn't even know their genders. Which leads me to the next important point...With an excellent mother plant, I consistently get a high quality **female** production plant, resulting in a more stabilized and superior harvest. Now that I have mothered the genetics I wish to grow, i.e. Bubba Kush and Super Skunk Kush, I will always have that same exact chronic bud every time I do a flowering cycle.

It's time to clear out the grow room of the male plants. It's easy, just bag'em and dump'em.

Less than three months ago I started with 10 seeds and now I have two excellent MOTHER PLANTS.

Place a plastic cup around the bottom of the plant to help prevent the main stem from becoming soggy.

Day 76. The mother plants are responding well to their new homes. They each have several good branches from which to take clones for flower production. It's good to remember never to cut too much away from the mother plant, as this can cause a plant to go into shock. It's not wise to take more than 30% of the green foliage away from any plant. This 1/2 pound grow only needs four production plants to fill the room so I'll cut six clones and flower the best ones.

Cloning

Production Plants.

Lets go over cloning one more time, this time for herb production! Start by selecting a healthy growing shoot. (Pic 1.) **Remember, healthy branches make healthy clones.** Go down the branch about 3-5 inches and remove the lower, smaller branches that are growing from the nodes. In the picture below, I'm showing how to make the final cut from the plant. With one hand I'm holding the branch, as I cut it from the plant at a 45-degree angle. I immediately dip the cutting into the Clonex™ cloning solution. Once the cutting has a generous amount of gel on its cut end, I then plop it into the 1 inch Oasis™ starting cube. I make sure to wet and label the cubes before the clones go into them. After I have finished cutting all the clones, I'll spray them down with water and place them under a humidity dome.

CLONING TIPS
*Use sharp, clean scissors.
*Cut @ 45⁰degree angles.
*Use a humidity dome.
-Try Clonex™(Cloning Solution).
-Try 1 inch Oasis cubes™.

FULLY LOADED VEGETATIVE ROOM

Both T5 fluorescents are providing light to the mother plants and the baby clones for 18 hours a day. It is best for the temperature of the rooms to be between 70^0 & 80^0 fahrenheit and for the humidity to be between 40% - 60%. The cuttings like 90% humidity until they grow roots, so the humidity dome really comes in handy. It's good to open up the dome and spray them down with water every couple of days. Waiting for the roots to grow may take up to a couple of weeks, but it usually happens by day ten. Once the roots pop through the cube, the clones will need more fresh air. Opening the dome lid for short periods of time will expose the cuttings to normal levels of humidity and helps lessen the shock. A few days after the roots are visible, the clones will no longer need the plastic dome. They will have grown a new root system and will be ready for the next phase of life.

DAY 86. Ten days have flown by and five of the clones have rooted. Now is the time to transport them into the 4 inch grodan™ cubes.

First, soak to "set" the Gro-dan® cubes in 5.5pH water. Next, shape the cube to fit a 1 inch square. Now, water the plants in the 1 inch cube with a mild nutrient mix of 500 ppm before fitting them into the pre-soaked 4 inch cube. (Photos 1, 2 & 3)

As the roots of the clones develop from the cubes, they will need more area to expand and grow.

The rockwool style Grodan® cubes are easy to work with and can be pinched to fit the square 1 inch Oasis™cube.

Getting the Grodan® Delta™cubes ready for the plant entails soaking the cubes in 5.5pH water for 30 minutes, in order to "SET" the cubes. Once the 3" x 4" cubes are "SET" simply plop the clone into the cubes 1 inch pre-cut hole and leave it alone. It's best not to water these plants for the first 3-4 days, as the slowly vanishing moisture promotes rapid root development.

Now is the time to trade out the little black tray with the 2'x2' white tray. I'll connect the drain hose and put down the catch-all tray on the floor. Place all the clones into the tray and pull the lights down to within 6 inches off the tops of the plants.

Roots! As the cube begins to dry-out over the next 3 to 4 days, the roots grow outward in search of water and nutrients. If the cube was watered every day, the roots would not have flourished as well. The key to the cubes is to let them dry-out a-bit in between waterings. Not too dry or the plants will die! Just to the point of the cubes being moist. There should still be some water weight left in the cube if I lift it up. If the cube is bone dry it will be very light to the touch. This would be bad, as the plant would be on the verge of wilting out and dying. **REMEMBER: From wet to moist before watering again.**

DAY 90. It's been three months since first germinating the Bubba Kush and Super Skunk Kush seeds. 14 days ago I took cuttings and they look great. The roots have grown through the 4 inch cubes and are looking for more area to grow. Let's introduce the 6 inch Grodan® cube. No tricks here, just remember to "set" the rockwool cubes as before (pages 52 & 65). I simply place the 4 inch cube directly on top of the 6 inch cube, but be warned as the top cubes will now dry out more quickly.

They will need water every day for a week or so until the roots grow into the 6 inch blocks. After the roots have grown for a few weeks, the block cubes will actually grow together and become one big block of roots. Try to be mindful when moving the plants around the room and remember to support both the blocks with both hands.

1000ppm Nutrient mix 1Gal
15-20ml Pro Grow™
10ml Karma™
5ml Sweet™
5ml Cannazym™
1ml Root Excelurator™
Adjust pH to 6.0-6.5

DAY 105. As the root system gets fully developed, the plants will turn their energy to foliar growth. Some farmers will flower the production plants at this stage and that is fine, but I want to increase the plants yielding potential. So, I'll vegetate these plants for another week or two.

The overall project has been working for four months,120 days. The clones are 6 weeks old and at 16 inches, they're now big enough to support some nice quality, chronic flowers.

TIME FOR FLOWER POWER!

Check out the 1" coco mat down on the 2 x2 tray. This is an important addition to the set-up because this mat will give the roots a place to grow beyond the block cubes. It is always wise to give the roots room to grow.

MOVING DAY 120. Moving the four production plants from the vegetative side to the flowering side of the growroom is easy. The 4 inch and 6 inch blocks have grown together. Try to be gentle with the plants when picking them up to avoid any sort of shock or discomfort. Once the plant transfer is complete, turn on the fans BUT not the light. The light stays off for the first 36 hours. This is a technique that is proven to trigger the cannabis plant to go into the flowering mode more quickly. So, try complete darkness for the first 36 hours, then 12 hours on & 12 hours off everyday thereafter.

This is a good time to flush the plants root system of any nutrient & salt build-up that may have accumulated. I'll run water with some Clear-ex™ through the base of the plants, and flush them out. Continue to flush water through the plants until the water coming out of the bottom of the block reads the same ppm and the same pH as the water going into the block. This is how to know that the plant has been cleaned of nutrient build-up. Once the plants have been flushed, I mix a batch of flowering nutrients for them. A 50/50 mix of both grow and flowering nutrients is good for the first week of flowering, because it'll provide a little more nitrogen (N) to support the plants rapid growth that takes place during the first week of flowering. The strength of the batch should be around 1200ppm. With the fans on and the light off, I'll roll down the door and leave the girls sleeping in the peaceful dark room for a day and a-half.

Notice the *flow* of the *grow*. From the mother plants come the production plants. They all grow-up under 18 hours of light in the vegetative growroom. Once the production plants reach the desired size, they are moved to the flowering room for flowering bud production.

1200ppm Nutrient mix 1Gal
10ml Pro Grow™ & 10ml Pro Bloom™
+10ml Karma™
+10ml Sweet™
+5ml Cannazym™
+5ml Ultra™
Adjust pH to 6.0-6.5

Although you may not be able to notice in this picture, I'm actually all fired-up because from this point on there is only eight weeks, 56 days, until we have our chronic cannabis flowers!

WEEK ONE FLOWERING

DAY ONE. The light is off for 36 hours with no interruption of the darkness. The plants must not be disturbed during this time. There is no need for watering, it's better just left alone. *The mother plants, on the other hand, on the other side, will continue to need care. The buckets should have the proper amount of nutrient water in them (approx. 1.5 gallons @1000ppm Veg grow mix) and the lights should be on for 18 hours a day to keep them in the summer mode.*

DAY TWO. The light is turned on for the first time in the flowering room. This picture holds true to the colors produced by the different lights. The T5 fluorescents are giving out the blue light on the right while the HPS is throwing out the orange glow. The HPS light will be producing a lot of heat and the six inch vent fan must be turned on to blast this hot air outside. It's important to keep the light as close to the plant tops as possible without burning them. I keep this 600watt light 16 inches away from the tops during the first two weeks, the "bud-set" time of growth. It is also important to try and keep the temperature of the flowering room between 70⁰ and 80⁰. I only have so many tools that are useful in controlling the environment that are available to me in this basic growroom. If we have a heat spell and it is 98⁰ in the shade outside, it is hard to keep the temperature down in this growroom without extra help from an air conditioning unit. The fact is, in some "hot" areas of the globe, it may be more wise to take the summer season off from the indoor grow project.

DAY THREE. The four plants are pictured here without the orange glow of the HPS light. Notice the dark green color and large size of the fan leaves. It has been a good 48 hours from the last time the plants were watered and the cubes are feeling just a bit moist to the touch, so today would be a good day to feed. I'll use a 5 gallon batch of nutrient mix and pour it over the cubes until they reach their saturation level and begin to release the excess water in the form of run-off.

Day four, five & six. This series of pictures demonstrates how the tips begin to cluster together as they first start to change from growing mode to flowering mode.

WEEK 1 NUTRIENT MIX 1200ppm
50%/50% TRANSITION MIXTURE OF THE BASE GROW AND BASE BLOOM NUTRIENTS.
1 GALLON CLEAN WATER.
10ML PRO GROW™ +10 ML PRO BLOOM™
+10 ML LIQUID KARMA™
+10ML SWEET™ +5ML ULTRA™
Adjust pH to 6.0 - 6.5
Weekly Clearex™flushing

DAY SEVEN. You can see how much the plants have grown when comparing this picture to the picture of day three on the opposite page. During the first two weeks of flowering the plants have an extreme growth spurt as they transition to the final phase of their life. Another factor for the rapid plant development is the 600watt HPS light which produces a lot more light than the T5s.

WEEK TWO FLOWERING

As we enter the second week of the flowering stage, **day eight**, the plants start to show their strong female traits as they throw stigmas out from the branch nodes. By **day nine** you can notice little clusters of flowers forming on the tops of the plants. The next two pictures, **day ten & eleven**, show just how the nodes and leaves start to cluster-up from below. All of these areas will grow into the fruit. The photo of **day twelve** shows just how big the four plants have gotten overall. Also, notice that the plants have grown rapidly but they did not stretch. Stretching is when the plant branches grow too quickly, pushing the buds farther apart from each other. When a plant stretches, it makes for spacey buds with a lower overall plant yield. Stretching can be lessened if the grower chooses to pay closer attention to the room's environmental details. The two main environmental factors that cause a plant to stretch are a lack of light and / or a cold night time room temperature. If the light is placed too high from the tops, the plants will react by stretching upward in search for more light. The best way to avoid this is to keep the light (600 watt hps) around 16 inches off the tops, during the first two weeks of the flowering cycle, when the plants are prone to stretching. After the first two weeks, the "bud-set" time, stretching tends to be less of a problem. The next factor is the night time room temperature, or better yet, the difference between the day and night time room temperatures. The greater the temperature difference, the more the plant tends to stretch. To minimize the plant stretching effect, try to raise the night time room temperature so that it is within 10 degrees or less of the day time room temperature.

WEEK 2 NUTRIENT MIX 1200ppm
100% BLOOM/ FLOWERING MIXTURE.
1 GALLON CLEAN WATER.
20ML PRO BLOOM™ + 10ML LIQUID KARMA™ + 10ML SWEET™ +10 ML BIG BUD™ + 5ML ULTRA™
Adjust pH to 6.0 - 6.5 *Weekly Clearex™flushing*

This means, in the winter season, it might be smart to bring a small room heater to warm-up the grow at night and lower the difference in temperature. This actual grow is taking place at the end of summer and I am running the 600 watt light during the cool evening hours and resting the plants during the warmer daylight hours. Remember the tricks employed to stop the plants from stretching: **Keep the light at 16 inches & warm up the nights (70°-75°). Day thirteen** is showing promise as the baby buds start to appear more developed. As the crop goes into **day fourteen**, I notice buds springing up everywhere.

Now is a good time to take clones for the next grow cycle. Remember that it can take up to six weeks to grow the cuttings up for flower production. So by starting the process now, I'll be ready for the next grow cycle as soon as this harvest is done.

WEEK THREE FLOWERING

During week three, the girl's energy shifts, from plant growth and "bud set", to actual bud development. The flowers should continue to grow everyday for the next four weeks! The nutrient mix I'm using is increasing 100ppm every week to feed the ever growing flowers. To avoid any potential problems of nutrient blockage, or root lock-up, I continue to flush the plants every week.

Day fifteen into flowering and the clusters are starting to develop quite well. **Day sixteen.** Almost overnight, the little cotton balls of love have taken hold all over the plants. Notice the leaves are a healthy, productive, dark indica green which is a good color for week three. *If the leaves on the plants look a little yellow, which sometimes can happen, I can add some **grow** nutrient to the flower mix. This will bring the nitrogen (N) levels up in the mix.

Once the leaves turn back to a healthly green color, I can go back to feeding them just the flowering nutrient mix. **Day seventeen** is slightly bigger than days 16 & 15. Everyday that I go check-up on my babies, I can observe the rapid growth. At this stage of flowering, the white stigmas in the bud are expanding in numbers and size.

DAY EIGHTEEN. Baby bud.

DAY NINETEEN. Solid dark green color in leaves.

DAY TWENTY. The buds are getting bigger every day.

WEEK 3 NUTRIENT MIX 1300ppm
100% BLOOM/FLOWERING MIXTURE
1 GALLON CLEAN WATER.
25ML PRO BLOOM™ + 10ML LIQUID KARMA™
+15ML SWEET™ + 10 ML BIG BUD™
Adjust pH to 6.0- 6.5 *Weekly Clearex™flushing*

Day 21. At the end of three weeks of flowering the plants have grown from 16 inches to 30 inches. As the plants complete the transition into flowering mode they will double in size. This will be the maximum height that these plants will grow. Now, raise the light a few more inches so that it's 18 inches from the tops of the plants. If the powerful light is too close it will burn the tops of the developing buds.

WEEK FOUR FLOWERING

As the first month of growth comes to a close the plants get heavier by the day. They are four weeks into full blown flowering mode and it's time to step-up the nutrient mix. Pay close attention not to make the mix too strong and damage the plants this close to the finish line! I have been slowly raising the PPM by 100 every week. Remember the importance of flushing the plants every week. Regardless of what may be planned on the grow schedule, if a plant shows signs of over-fertilizing or stress, FLUSH, FLUSH, FLUSH! Some of the first signs of over-fertilization is leaf tip burn and curl. When in doubt, flush. The flushing is complete once the water run-off draining from the plants reads the same PPM and pH as the water being poured into the plants. This week I'm pushing the plants to 1400ppm. Almost maximum. As far as the rest of the room is concerned, all that is needed is to make sure that the light turns on and off and that the fans keep moving the air around. In this series of photos, the buds are growing a little bit every day and there is some trichome crystal development forming too. In the first part of the week, the buds seem to be gaining more overall girth and by the end of the week, it is obvious that the buds are connecting together and swelling-up.

Just look at the number of white stigmas growing out from the calyx pods in the photo of **day 25**.

WEEK 4 NUTRIENT MIX 1400ppm
100% BLOOM/FLOWERING MIXTURE.
1 GALLON CLEAN WATER.
25ML PRO BLOOM™
+10ML LIQUID KARMA™
+20 ML SWEET™
+10 ML BIG BUD™
+10 ML OOZE™
Adjust pH 6.0- 6.5
Weekly Clearex™flushing

Day 26 The buds from the lower part of the branch are starting to connect with the buds growing from the top, forming one big fat bud! The THC filled "fuzz", trichome crystals, are really starting to show themselves by **day 27.** These crystals are forming all over the bud in an attempt to get some male pollen to stick to it. This is where the story gets even more interesting. You see, the longer the buds grow without being pollinated, the more the THC resin fills the crystals in an attempt to get male pollen to bond with her flower. The trick for the grower is to get the plant to produce as much potent THC crystals as possible without getting the plant pregnant with seeds. That should not be a problem because there are no male plants in this garden and thus no male pollen. Therefore, there are no seeds! This is a top view of a bud on **day 28,** halfway done!

WEEK FIVE FLOWERING

During the next two weeks, the flowers explode with growth and trichome crystal development. It is normal for the buds to start maturing in color around this time. Notice how the white stigmas are slowly changing colors to amber and orange/brown.

This is a sign to the grower that the flowers are starting to ripen. The daily pictures tell the story. **Day 30** and **day 34** are pictures of the same exact flower, four days apart. You can see the stigmas starting to change colors.

WEEK 5 NUTRIENT MIX 1500ppm
100% BLOOM/FLOWERING MIXTURE.
1 GALLON CLEAN WATER.
30 ML PRO SOIL BLOOM™
+10 ML LIQUID KARMA™
+20 ML SWEET™
+10 ML OVERDRIVE™
+15 ML OOZE™
Adjust pH to 6.0- 6.5
Weekly Clearex™flushing

Day 35 and this flower is looking awesome. The bud has connected from the bottom up and the crystals are everywhere. These flowers are starting to look so juicy!

WEEK SIX FLOWERING

Week six is the last chance for the buds to put on any real weight. After this week, it's all about crystal development and ripening. This is also the last week of using a high dosage of nutrients at 1500ppm. The buds will continue to swell throughout the week of growth.

DAY 36

DAY 39

DAY 37

DAY 40

DAY 38

DAY 41

Look at the buds in the picture below. Most beginner growers would look at these buds and think they are ready for harvest. The fact is this bud still has a week to go before it's perfect for harvest. It looks ready to the untrained eye because it is big, sticky and stinky! BUT there is still more crystal development underway and it would be a shame to harvest now so close to peak perfection. Remember, these are indica kush plants with an expected flowering life of eight weeks; 56 days. It's common for the plants to finish a couple of days early or a little late, depending on the overall health of the plants and average temperature of the growroom. A warm room can finish the plants a little early and a cool room can make them mature later.

WEEK 6 NUTRIENT MIX 1500ppm
100% BLOOM/FLOWERING MIXTURE.
1 GALLON CLEAN WATER.
30ML PRO SOIL BLOOM™
+10ML LIQUID KARMA™ +20ML SWEET™
+10ML OVERDRIVE™ +15ML OOZE™
Adjust pH to 6.0- 6.5
Weekly Clearex™flushing

WEEK SEVEN FLOWERING

We are entering the final phase of flowering and the focus has now turned from bud growth to crystal production and maturing. Now is the time to try and make any last minute adjustments to the indoor environment. All the adjustments are meant to signal to the plant that the end is near. First, I raise the light height. Instead of 16-18 inches for the 600watt I'll raise it to 24 inches or more. I'm trying to mimic the fall season when the earth tilts away from the sun. This triggers the plants to ripen-up quicker. I also feed the girls with cold water, in the low 50^{0}s. Another thing I do is cut down the hours of light per day, from 12 hours to 10. These tricks help trigger the plants to go into the ripening mode, due to the oncoming winter season.

Wow, I must say these girls are really starting to look 100% ripe. It's time to bust out the magnifying lens to get an up-close view. First, I'm looking to see that at least 70% of the stigmas have matured, changing from the white color to the more matured orange/brown color. Second and more importantly, I am closely examining the trichome resin crystals to see if they are at maximum potency. I'm looking at the whole trichome. When peaking, they stand up tall and are capped with a little resin ball. This ball should look full of resin and ready to pop. The resin inside should be a clearish, light milky white color. Once the crystals start to turn an amber color and begin to bend and break off, the bud is moving past its peak.

WEEK 7 NUTRIENT MIX 700ppm
50% BLOOM/FLOWERING MIXTURE.
1 GALLON CLEAN WATER.
10ML PRO SOIL BLOOM™
+ 5ML LIQUID KARMA™
+ 10ML SWEET™ + 5ML CANNAZYM™
+5ML OVERDRIVE™
Adjust pH to 6.0- 6.5
I started flushing with Clearex™ on day 47

WEEK EIGHT

Flush, flush, flush! It's a good idea to start flushing at least seven days before the harvest. It's possible to flush out plants with less time but I recommend at least seven days because I WANT TO FLUSH OUT ALL THE NUTRIENTS BEFORE THE HARVEST. Begin by flushing a gallon, or two, of Clearex™ mix through each plant everyday, for 3-4 days. Remember, the way to check that the plants are being properly flushed is to examine the water draining out from the plants. When the runoff water reads the same ppm (+ or -100) as the clean water being pumped in, then the plant has been flushed good enough. After a few days of that treatment, I'll switch to just plain water (0.0ppm - distilled if available) and flush for the remaining days until harvest. During this week I'll raise the light to 30 inches off the tops and have it ON for 9 hours per day. I also try to keep the room's air temp as cool as possible, 60's during the day, 50's at night, to simulate the naturally cooler fall outdoor temperatures.

WEEK 8 FLUSHING MIX 00-150ppm
100% FLUSHING MIXTURE.
1 GALLON CLEAN WATER +30ml Clearex™
Adjust pH to 6.0- 6.5
Daily Clearex™ flushing
WATER ONLY FLUSH LAST 2-3 DAYS

Plants do not always finish ripening at exactly 8 weeks. They are living things. The grower must use his/her judgement to determine when the plant is at 100% peak potency and this can give or take a day or two or three... Perfectly ripened crystals can be seen in the picture of day 52, as this bud is ready for picking.

Day 54, 48 hours shy from the completing of the entire eight week flowering cycle. After a close examination of the resin filled crystals, I believe that THIS CROP IS READY FOR HARVEST RIGHT NOW! The milky, translucent crystals are standing straight up and filled with THC. Once the decision is made to harvest the plants, try turning off the light and giving them 10 hours of sleep. With the light off, the plants stop photosynthesis and the sugars and carbohydrates

inside the plant begin to retreat back downward into the root system. Good farmers try to take advantage of this natural occurrence. These sugars and carbohydrates are essential for the plant's health during its life cycle, BUT they can make the buds taste harsh and grassy if not flushed properly from the plant. To avoid that homegrown grass taste, it's important to flush the plants out during the last week. AND try harvesting the buds "in the DARK", while they sleep. I'm not totally in the dark. I use a green light to see in the flowering room when the 600 watt HPS is turned off. The green light does not disturb the sleeping plants. This is due to the fact that the green spectrum of light is reflected, not absorbed, by the plants. Thus, their green color. To harvest the plant, cut the main stock away from the root system. If growing different strains of plants, it's a good idea to keep track of what plant is where, as they can all begin to look alike after they are cut up and laying in a pile.

Stalks can get much bigger than this, but that would require more time growing in the vegetative state. For a growroom project this size, a 1/2 inch to 1 inch stalk, at the time of harvest, is fine. That size of plant stalk will easily be able to support 1-2 ounces of flower.

"Thank you ladies for a job well done." Check out the plant's root blocks after harvest. Notice that the 4 inch cubes placed on the 6 inch cubes have grown together completely as one.

This is why I like to use the coco mat under the grow cubes. Just look at how the roots have grown into the coco mat. It's important to remember that cannabis roots need ample space to grow. If we limit the size of the root zone we limit the size of our harvest! The coco mat provides space for the roots to expand, without growth restrictions.

Now, onto the trimming and hanging part of this story. Nobody wants to mess this part up. The main thing to avoid is mold. Mold can grow and spread very quickly and can ruin all of the herb. The harvest area needs be clean, dry and have good air flow. The buds should hang and cure in a dark area as light can alter and damage the resin filled flowers during the curing and storage process. I start trimming by cutting off a branch and slowly twisting it around cutting off all of the excess leaves growing around the buds. I'm trimming my bud up tight. There are a few different trim styles from which to choose: country or city. A 'country' cut means to cut off all the big and medium leaves while leaving the smaller, crystally leaves on the bud, as protection. The 'city' trim takes more time to finish as it entails removing ALL of the leaves from the flowers. These flowers are trimmed up tight because I want them to bling and the 'Chronic' style of bud is traditionally

trimmed in this manner. This works out because I like to save the crystal covered leaf to use for making kief. Or try using the chronic trim for baking butter. Or use the trim of chronic leaf schwag to mix-up a batch of bubble hash! After trimming down a branch, hang it in the curing box (cardboard box and fishing line). I keep trimming the branches of the plant until the whole thing has been cleaned and is hanging. Then I do the same for the other three plants making sure not to confuse the Bubba with the Super Skunk.

REPEAT

The harvest is all manicured and hanging up in my home-made curing box. It is important to place the box in a dark, well ventilated room. It's also a good idea to use a carbon filter in the drying room, during this stage of the grow. It can tend to be the strongest smelling period of the entire grow cycle! After three or four days, start to check the buds for signs of dryness. I find that it is best for the overall taste and quality when the flowers dry out slow and evenly, over the course of 7 to 14 days. The fact is, unless a grower has an environmentally controlled drying room the harvest will be subject to climate fluxuations. Meaning that... if there is a heat wave, maybe the bud will dry up in four days; If there is a wet rain storm outside, maybe it will take 21 days to dry. Keep the air moving. Remember, it is always possible to remoisten bud if it gets a little too dry, but it is impossible to un-mold bud. So, for best results, keep the drying room dark and dry, and keep the air moving. We'll come back to this hanging harvest in a week or so to jar up the buds and begin the sweating (curing) process.

CLEANUP AND RELOAD

CYCLE

While the bud is drying it's a good time to clean up the grow-room and reload it with four new plants. The production clones have been quietly growing up in the vegetative room for the past six weeks. Just wait and see how quick and easy it is to re-boot this system and start producion on the next crop! Before I get too excited, here are some tips on cleaning:

-- Throw away all of the old trash and plant matter.

-- Use a towel, wet with a mild bleach solution to wipe down everything!

-- Clean the grow tray and table.

-- Clean the walls, floors and door.

-- Clean the glass on the light.

-- Clean it all.

-- Place a new coco mat in the tray.

I don't suggest re-using the coco mat, nor the cubes, as this may lead to plant disease and crop failure.

CURING THE FLOWERS

As stated, the drying process is a very important part of the job. And the job is not complete until the flowers are dried, cured and stored safely in an airtight jar. But how can you know for sure when the buds are totally dry? Well, there are two tests that I use to determine the moisture of the buds before I start to jar them. The problem with jarring up wet bud is that it can mold! On the other side of the spectrum, there is a point at which the buds can become so dry they risk crunching apart when touched. I'm looking for something right in the middle, the sweet spot.

#1 The pinch. At first, I'm looking for the bud to crunch a little bit on the outside when pinched. It still needs to have some moisture at its center. At this point, the bud is not ready to smoke. I have to pull the moisture out from the center, the core of the bud. The flowers need to be dried evenly throughout and cured properly before they can be smoked. *Try and refrain from smoking 'wet bud' as this can be very unhealthy and can cause a nasty sickness, commonly known as 'wet lung'. Not nice.*

#2 The stem snap. I'm looking for the branch to break, not just bend, but break. I'm looking to get at least a good crunch or a snap sound when breaking a branch. Once the flowers pass both of these tests, it's time to jar them up and start the sweating process.

I use some clean, wide mouth, mason jam jars, available at most supermarkets to keep the buds. It's important for the jars to be able to close airtight. I'll snip the buds off the branch and put them into the jar one-by-one. Once the jar is full of bud, I'll close the lid and let the sweating process begin... Everyday opening the jars to let the buds air-out. Over the next week, I'll check the buds in the jars as the moisture from deep inside the flowers is pulled outward. This is what curing is all about! The bud re-moistens when the lid is closed and releases that moisture when the lid is off. This is the correct process for an even cure of the cannabis flowers; 'sweating it out'.

AS

Five jars total

2 Jars Super Skunk Kush = 3.5 ozs
3 Jars Bubba Kush = 4.5 ozs
8 ounces of chronic!
That's one half-pound of chronic!

More on curing the bud: I open the jars once or twice a day for 10 to 20 minutes to release any moisture and let the buds air out. This is the last part of the curing process. What's happening is that the moisture from deep inside the bud is being drawn out, towards the more dry, outside part of the bud. That is why the semi-crunchy, dryish bud, that I have put into the jar, becomes re-moistened and perfect again overnight. I want to avoid putting wet bud into the jar because it can become moldy. So, if I discover the buds are a little wet, I'll leave the lid off of the jar for a few hours so that the buds dry out before resealing the jar. This is what great curing is all about... Slow drying. Re-moistening, airing out, remoistening, airing out. The buds get better with time. If done correctly, it can take up to a month for the flowers to be cured perfectly. This is why the job was not actually finished when I first jarred-up the bud. The flowers were dry on the outside, but still damp on the inside. Only after a couple of weeks of working with the harvested flowers in this manner, can I write with confidence, that they are perfect for blazing.

What if the bud gets too dry?

NEEDED

If the harvested flowers get over dry, like "dust-crunchy" to the touch, I can always try to remoisten them. I do this by adding a fresh leaf or two to the jar for a 24 hour period. After the buds have remoistened, remove the leaf and reseal the jar. This usually does the trick, with stellar results.

Once the flowers have been completely dried and cured it is time for storage. I keep the buds in jars and make sure to store them in a cool, dark place. I air-out the jars once a week, or as needed, to freshen them up. The buds can be stored in this manner for three to four months without losing potency. For long term herbal storage techniques visit: www.igrowchronic.com

Some homegrown

CHRONIC +

The original Jerome Baker

SUN BONG

***from the movie* +**

GRINDER +

KIEF TOPPER =

"So now you know."

THE END

Here are some basic graphs showing the preferred nutrient levels, in parts per million (PPM), for the entire lifecycle of a cannabis indica dominant plant. If working with a sativa hybrid, carry the high PPM of week 5 and 6 in flowering through to week 7 of the flowering cycle then pull back hard for the flush.

VEGETATIVE NUTRIENT FEEDING SCHEDULE

This graph includes starting the plants from seed (week 1), to sprouts (week 2-3), to young plants (week 4 thru 8) and on to fully mature plants (week 9 -10+).

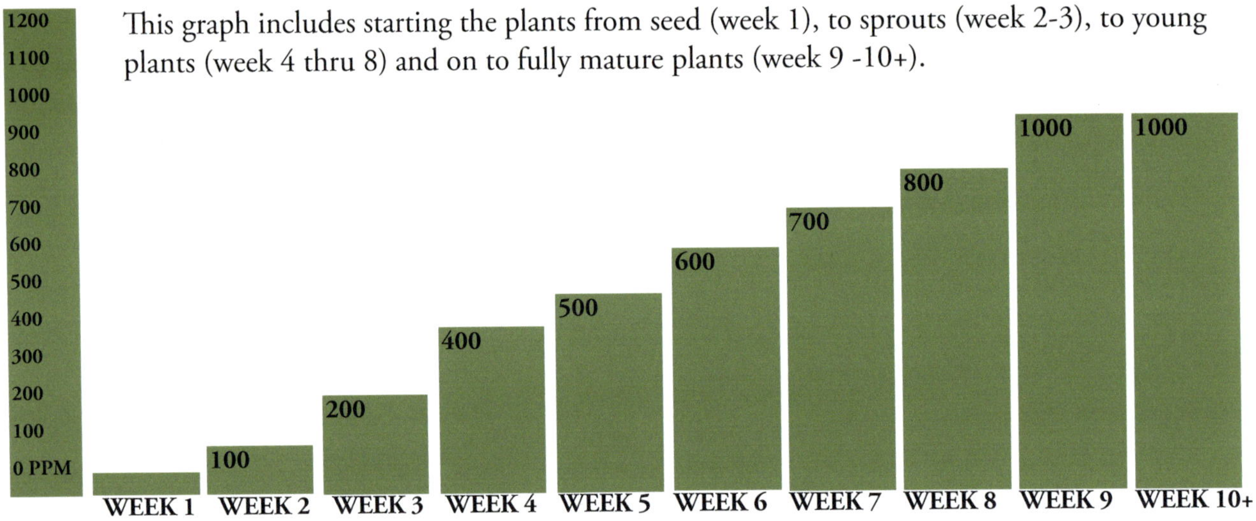

FLOWERING NUTRIENT FEEDING SCHEDULE

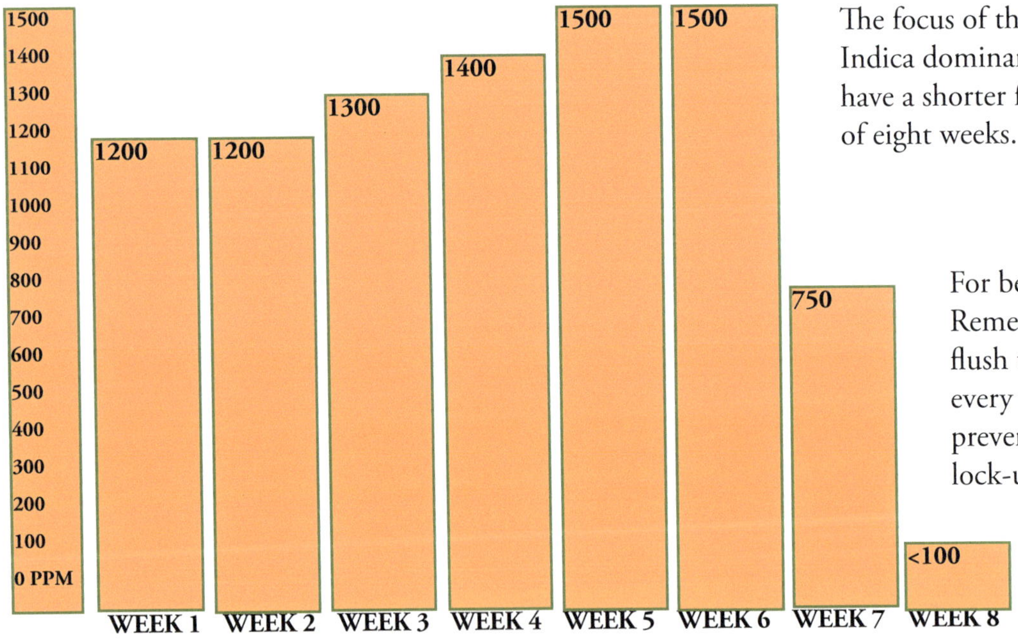

The focus of this graph is on Indica dominant plants that have a shorter flowering time of eight weeks.

For best results... Remember to flush the plants every week to help prevent nutrient lock-up.

NOTES:

SHOPPING LIST:

SOME BASIC CONSTRUCTION MATERIALS & TOOLS:
ELECTRIC DRILL; DRILL BITS; SCREW DRIVER BIT; ELECTRIC WOOD SAW; TAPE MEASURE; DUCT TAPE; DRYWALL SAW; STAPLE GUN; STAPLES; VELCRO.

MATERALS:
10x's 8' wood 2X4's; box of 1" & 3" drywall or wood screws; 3 or 4 L-brackets for support; drywall; 4" & 6" vent connectors. Black Plastic & Cardboard Cutouts.

GROWROOM EQUIPMENT:

1. T5 fluorescents for veg room. 2. Small carbon filter. 3. Four inch vent fan. 4. Four inch vent tube. 5. Two 2'x2' grow trays. 6. Waste water trays. 7. Hydro-buckets™ for the mother plants. 8. Two small room fans for veg room & flowering room. 9. Closed, reflective hood for bud room. 10. HPS electric ballast. 11. 50cm carbon filter. 12. Six inch vent fan. 13. Six inch air vent tube.

NUTRIENT LIST:

-PURE BLEND-PRO GROW™ By Botanicare®.
-PURE BLEND-PRO BLOOM™ By Botanicare®.
-PURE BLEND-SOIL MIX-PRO BLOOM™ By Botanicare®.
-SWEET™ By Botanicare®.
-LIQUID KARMA™ By Botanicare®.
-ROOTS EXCELURATOR™ By House & Garden Van DeZwaah®.
-GREAT WHITE™ By Plant Revolution®.
-BIG BUD™ By Advanced Nutrients®.
-OVERDRIVE™ By Advanced Nutrients®.
-CANNAZYM™ By Canna®.
-pH-DOWN™ By Advanced Nutrients®.

I NO LONGER USE THESE PRODUCTS *(BUT DID FOR THIS GROW).*

-BLOOM, ROOTS™ By Agricultural Organics®.
-BLOOM, PRE™ By Agricultural Organics®.
-BLOOM ULTRA™ By Agricultural Organics®.
-BLOOM OOZE™ By Agricultural Organics®.

INSTEAD TRY:

-HYDROPLEX™ By Botanicare®.
-TERPINATOR®. Botanical Plant Oil.

(These are the products I used for this grow. They are exactly what was used for the book with the idea being "if I got these excellent results using these products, then so could anybody interested in quality." They are all good products BUT they **all** can be substituted for similiar products from other companies. There are many companies and products from which to choose. Your local grow store can help you select these or other products.)

BAD BUGS are a natural part of farming and they are sure to provide many challenges to all of us growers, even the cleanest, most organized of us. It does really help to have a clean growroom to fight off and keep out bugs. It also really helps not to bring infected plants, from an outside source, into your growroom! This is, by-far, the number one way to infect your garden. But the fact remains, bugs are hungry, tiny little monsters and they can find many ways to get inside the garden. Once they arrive, they will start sucking the life juice from your plants.

EVERY FARMER WILL, AT ONE TIME OR ANOTHER, HAVE BUG ISSUES...

It's how you deal with the bug problems, in your garden, that separates the great farmers from the average ones. A great farmer never uses pesticides or fungicides of any sort. A great farmer finds better ways to go about their bug killing. Most chemicals used in generic pesticides are harmful to us, the plants and the earth. It's important for the health of your family and the health of the planet to learn better ways to "control nature". There are other "natural" pesticides on the market, such as "neem oil" or "pyrethrum spray" but I must warn you that these types of "natural" sprays have killed my plants in the past. If you mix too strong of a dose, your plants will burn-up and die. Test a small area of one plant first and watch it for 24 hours before you dose the whole garden. Use at your own risk.

When called into action, I spray the plants with soapy water to fight off a bug infestation that needs immediate attention. The problem with soapy water is that it is not very effective at killing off all of the bad bugs. It does, however, keep the infestation down while I'm waiting for a shipment of predator bugs! FOR EVERY BAD BUG, THERE IS A GOOD BUG. You never have to poison your crop with chemicals, if you can get the good bugs to eat all of the bad bugs. There is not a pest that is damaging to the cannabis plant that does not have a natural predator! Even the most damaging of all the bad bugs, the spider mite, has a natural predator.

In my garden, I use a pro-active approach in my defense against bad bugs. I employ a strong offense of predator bugs to roam the garden and eat anything before it ever has a chance to populate in my garden. My private army consists of ladybugs, pirate bugs, green lacewings and predator nematodes. All of these good bugs work hard to rid my garden of the bad bugs before they can take a bite out of my herb!

Many bad bugs are strongly attracted to the color yellow! These include fungus gnats and whiteflies. I'll use sticky yellow traps just to see what's going on in the garden and when fighting infestations.

GOOD BUGS

SPECIFIC PEST CONTROLS

-WHITEFLY PARASITES
-WHITEFLY PREDATORS

KILL, KILL, KILL!

-LADYBUGS
-GREEN LACEWINGS
-APHID PREDATORS
-APHID PARASITES

DIE, DIE, DIE!

-PREDATOR NEMATODE
-THRIP PREDATOR MITES
-PIRATE BUGS
-STICKY BLUE TRAPS

KILL, KILL, KILL!

-FUNGUS GNAT PREDATOR
-PREDATOR NEMATODE
-STICKY YELLOW TRAPS

DIE, DIE, DIE!

-MEALYBUG DESTROYERS

KILL, KILL, KILL!

-SPIDER MITE PREDATORS
-MITE DESTROYERS
-FELTIELLA ACARISUGA

DIE, DIE, DIE!

PRO-ACTIVE PEST CONTROL

-LADYBUGS
-PIRATE BUGS
-GREEN LACEWING LARVAE
-PREDATOR NEMATODES

Lady bugs are a pro-active approach to keeping many bad bugs at bay in the garden.

BAD BUGS

-WHITEFLIES

...are small, evil, white bugs. They fly around sucking and their pupae (babies) suck, too. They hang out on the backside of the leaves and crap out a shiny honeydew that forms into mold and can kill the plant.

-APHIDS

...come in all colors, they are soft bodied and shaped like a pear. Evil aphids have two "horns" growing out of their backside. These small bugs like to suck the young growing tips of the plant.

-THRIPS

...are tiny and move fast! I notice the damage they cause, before I notice them. They cut, then suck, the juice from the leaves of the plant, leaving a shiny scar. They raise their youth in the roots and move on to the leaf for a life of munching on my chron. Bad.

-FUNGUS GNATS

...look like whiteflies without the white color. They can cause small amounts of root damage to young plants, but they really are more of a nuisance, than a problem. Not as evil a bug, but not really likeable either.

-MEALYBUGS

...look like little drugged-up ladybugs, with a white ghost blanket pulled over their heads. They like to cluster around the joints of the plant ..Evil, but slow moving and slow populating, so manageable.

-SPIDER MITES

...are truly the most evil, the most damaging of all these bad bugs. They are tiny, move quickly around the garden and breed rapidly! Most often, you will find the tan/greenish mites, infested in a fine webbing that covers the tops of your plants. It is as if they appear from nowhere and destroy within days! Bag and remove infected parts of the plant from the garden immediately and get predator bugs shipped overnight. After the harvest, make sure to really clean the growrooms and consider bug bombing the area. If your mother plants in the vegetative side have been infected too, then take the bug battle over to them. You will have to fight them off hard, or else consider getting rid of everything and starting over from seed.

The tiny thrip bug scrapes and sucks the juice from the plant, leaving scars on the leaf. These pests suck, but can be controlled with pirate bugs, nematodes, predator mites and sticky blue traps! Fight, fight, fight!

SOME INTERESTING GROW PHOTOS

This is the young tap root growing from my Bubba Kush Mother plant. The roots are growing out of the holes in the bottom of the holding bucket.

The leaves of the cannabis plant will change from a dark green color to the colors of fall season. It's natural and normal for the leaves of the plants to change colors to yellow, purple and brown as the harvest day nears.

This is what mold looks like. If any part of a plant or bud gets moldy it's important to cut that part away and remove it from the garden, before it has a chance to spread. To avoid this potential problem, keep the humidity in the room down. Lower levels between 40% and 50% are good. If you find yourself battling mold and high humidity, try running a dehumidifier in the room.

GLOSSARY

Air-o-ponic- Type of hydroponic watering system that delivers oxygenated water and nutrients to the plants root system.
Bud Set- The time during the first half of the flowering cycle when the plant begins to develop buds.
Budding- When the cannabis plant develops and ripens its flowers. 12 hours of light.
Calyx- Is a small pod growing on the female plant that will have a stigma (pistil) coming out of it. If pollinated, this is where a seed will grow.
CBD- Cannabidiol is a major ingredient of cannabis that is in high demand because of its many potential medical uses.
Clones- Cuttings from a plant that develop a root system and become plants of their own. A clone will carry the same genetic code as its parent.
Curing- The drying process. The act of pulling the moisture out of the flowers.
Diff- Difference. Refers to the difference in daytime and nighttime temperatures.
Flowering- Budding time! When the plant grows and ripens its buds. 12 hours of light.
Flush- Washing the root system of nutrient build-up.
Herb- Cannabis, ganja, pot, grass, weed, the chronic.
Hydroponic- Grown without soil.
Internodal- The space inbetween growing branches or nodes.
Nodes- The joints of the branches and the main stem.
Nutrients- The juice that you mix with water to feed the plants.
PGR's- Plant Growth Regulators. Plant hormones. Bad stuff.
PPM- Parts Per Million. Unit of measurement used to determine the amount of nutrients in the water mix.
Pupae- Baby bugs.
Stretching- When a plant grows really fast searching for light. Long, undesired internodal growth.
Stigma- The white hairs that turn brown towards harvest time. They grow from out of the female calyx pod.
Sweating- Part of the curing and drying process referring to the act of pulling the moisture from deep inside the bud, outward.
THC- is the main psychoactive ingredient of the cannabis plant. Short for Tetrahydrocannabinol.
Trichome- The crystal in which the prized cannabis resin is stored.
Vegetative- This is the 'summer mode' when the plants do their major growing. 18 hours of light.

Made in the USA
Las Vegas, NV
03 November 2026

57423044R00069